AF866223

www.KnowledgePublications.com

SAVE HARMLESS AGREEMENT

Because use of the information, instructions and materials discussed and shown in this book, document, electronic publication or other form of media is beyond our control, the purchaser or user agrees, without reservation to save Knowledge Publications Corporation, its agents, distributors, resellers, consultants, promoters, advisers or employees harmless from any and all claims, demands, actions, debts, liabilities, judgments, costs and attorney's fees arising out of, claimed on account of, or in any manner predicated upon loss of or damage to property, and injuries to or the death of any and all persons whatsoever, occurring in connection with or in any way incidental to or arising out of the purchase, sale, viewing, transfer of information and/or use of any and all property or information or in any manner caused or contributed to by the purchaser or the user or the viewer, their agents, servants, pets or employees, while in, upon, or about the sale or viewing or transfer of knowledge or use site on which the property is sold, offered for sale, or where the property or materials are used or viewed, discussed, communicated, or while going to or departing from such areas.

Laboratory work, scientific experiment, working with hydrogen, high temperatures, combustion gases as well as general chemistry with acids, bases and reactions and/or pressure vessels can be EXTREMELY DANGEROUS to use and possess or to be in the general vicinity of. To experiment with such methods and materials should be done ONLY by qualified and knowledgeable persons well versed and equipped to fabricate, handle, use and or store such materials. Inexperienced persons should first enlist the help of an experienced chemist, scientist or engineer before any activity thereof with such chemicals, methods and knowledge discussed in this media and other material distributed by KnowledgePublications Corporation or its agents. Be sure you know the laws, regulations and codes, local, county, state and federal, regarding the experimentation, construction and or use and storage of any equipment and or chemicals BEFORE you start. Safety must be practiced at all times. Users accept full responsibility and any and all liabilities associated in any way with the purchase and or use and viewing and communications of knowledge, information, methods and materials in this media.

www.KnowledgePublications.com

MANUALS OF CHEMICAL TECHNOLOGY—VII

INDUSTRIAL GASES

INCLUDING

THE LIQUEFACTION OF GASES

AND THE MANUFACTURE OF

Hydrogen, Oxygen, Nitrogen, Carbon Dioxide, Sulphur Dioxide, Ammonia, Producer Gas, Illuminating Gas, Acetylene, Ozone, Etc. Etc.

BY

GEOFFREY MARTIN, D.Sc., Ph.D., F.C.S.; ERNEST A. DANCASTER, B.Sc.; J. M. DICKSON, B.Sc.; FRANK B. GATEHOUSE, F.C.S.; E. JOBLING, A.R.C.Sc., B.Sc., F.C.S.; and H. STANLEY REDGROVE, B.Sc., F.C.S.

NEW YORK
D. APPLETON AND COMPANY

PREFACE

THE present volume forms the seventh of our series of manuals of technological chemistry, and deals with the subject of gases of industrial importance.

The subject of industrial gases has been revolutionised within recent years. Perhaps the most striking advances have been made in regard to the subject of the liquefaction of gases, especially of air. As a result of such advances, large industries concerned with the production of gaseous oxygen and nitrogen have come into being within the last few years, industries which have already, to some extent, revolutionised engineering practice and which have created entirely new industries such as the manufacture of cyanamide, now so widely used, as a manure and as the basis of manufacture of many nitrogen compounds.

The subject of hydrogen gas has quite recently acquired great importance, partly on account of the enormous quantities required for the new processes for hardening fats, and partly for military purposes for filling airships. An enormous amount of recent research work has been done on this subject, and methods have been worked out for producing the gas cheaply on the large scale which were quite unknown a few years ago. A full account of these new methods is given in the following pages.

Great changes have likewise occurred within the last few years as regards new methods of producing ammonia, and synthetic ammonia is now an accomplished fact. Of course, full details of the actual methods employed in this connection are carefully guarded as valuable trade secrets. What is known of the subject, however, is given as fully as possible in the following pages.

The last chapter is concerned with industrial ozone, a gas whose importance is gradually increasing as its uses become more widely known.

In this volume we have collected together a great many facts which have never before appeared in book form. Chemical engineers, chemists, and other technical experts will find in these pages a lot of data which previously were either hidden away in the often inaccessible pages of a scattered chemical literature or buried away in patents. The patent literature has been made as complete as possible, since many patents, although impracticable, often contain the germs of valuable ideas.

It is hoped, therefore, that a wide circle of readers will find the present volume of assistance as a handy work of reference.

The editor will be much obliged if practical men will call his attention to any errors or omissions in the present work which have escaped his attention.

GEOFFREY MARTIN.

TABLE OF CONTENTS

CHAPTER I

THE LIQUEFACTION OF GASES, INCLUDING THE MANUFACTURE OF OXYGEN, NITROGEN, AND HYDROGEN FROM LIQUEFIED GASES

By JOHN M. DICKSON, B.Sc. (Lond.), Chemical Engineer

CHAPTER II

INDUSTRIAL OXYGEN

By GEOFFREY MARTIN, D.Sc., Ph.D.

CHAPTER III

INDUSTRIAL NITROGEN

By GEOFFREY MARTIN, D.Sc., Ph.D.

CHAPTER IV

HYDROGEN

By H. Stanley Redgrove, B.Sc. (Lond.), F.C.S.

CHAPTER V

PRODUCER-GAS

By H. Stanley Redgrove, B.Sc. (Lond.), F.C.S.

CHAPTER VI

THE CARBON DIOXIDE (CARBONIC ACID) INDUSTRY

By H. Stanley Redgrove, B.Sc. (Lond.), F.C.S.

CHAPTER VII

MANUFACTURE OF NITROUS OXIDE

By Geoffrey Martin, Ph.D., D.Sc.

CHAPTER VIII

THE AMMONIA AND AMMONIUM SALTS INDUSTRY

By Geoffrey Martin, Ph.D., D.Sc.

CHAPTER IX

MANUFACTURE OF SULPHUR DIOXIDE AND SULPHITES

By Geoffrey Martin, D.Sc., Ph.D.

CHAPTER X

ACETYLENE

By Frank B. Gatehouse, F.C.S.

CHAPTER XI

THE ILLUMINATING GAS INDUSTRY

By Ernest A. Dancaster, B.Sc.

CHAPTER XII

INDUSTRIAL OZONE

By E. Jobling, A.R.C.Sc., B.Sc., F.C.S.

CHAPTER I

The Liquefaction of Gases

CHAPTER I

THE LIQUEFACTION OF GASES,

Including the Manufacture of Oxygen, Nitrogen, and Hydrogen from Liquefied Gases

JOHN M. DICKSON, B.Sc.(Lond.), *Chemical Engineer*

THE following are the chief patents :—

English.—*Aumont*, 11,306, 1911. *Bobrick*, 25,829, 1908. *Claude*, also *Soc. L'air Liquide*, 12,905, 1900; 28,682, 1903; 26,435, 1905; 27,658, 1902; 29,733, 1906; 12,358, 1904; 17,216, 1909; 7,305, 1910; 16,298, 1903; 20,349, 1907; 7,175, 1910; 3,302, 1907; 3,326, 1911; 22,316, 1909; 11,710, 1907; 5,395, 1909. *Cases*, 18,386, 1911. *Code and Knudsen*, 19,838, 1901. *Dumars*, 11,126, 1903; *Flüssige Luft Heylandt*, 16,615, 1909. *Goldschmidt*, 12,743, 1908; *Hampson*, 10,165, 1895; 7,559, 1896; 7,773, 1898. *Hazard and Flamand*, 26,720, 1908; 27,592, 1908; *Hers*, 21,191, 1901. *Hildebrandt*, 11,212, 1905; 13,790, 1907; 25,517, 1907; 6,515, 1909; 13,034, 1907; 26,186, 1907; 3,316, 1910; 15,999, 1910; 19,336, 1910; 1,828, 1907. *Jänecke*, 23,107, 1909. *Joly*, 21,861, 1899; 12,918, 1901; 15,511, 1901. *Knudsen*, 23,646, 1902; 6,087, 1903; 19,033, 1906. *Lake*, 11,609, 1902. *Leptien*, 2,497, 1910. *Leslie*, 11,902, 1906. *Levy*, 16,615, 1902; 5,649, 1903. *Levi and Hellbronner*, 20,053, 1902. *Linde*, also *Ges. für Linde Eismachinen*, 12,528, 1895; 14,111, 1902; 11,221, 1903; 7,205, 1911; 9,260, 1911. *Lilienfield*, 22,930, 1911. *Loumet*, 17,605, 1902. *Marks*, 25,829, 1908. *Merves*, 24,144, 1905; 21,780, 1906. *Ostergreen and Burger*, 26,361, 1898. *Parkinson*, 4,411, 1892. *Pictet*, 14,303, 1903; 2,713, 1901; 19,254, 1900; 21,120, 1902; 14,431, 1904; 27,463, 1910. *Place*, 11,609, 1902; 13,529, 1907; 3,685, 1908. *Ramsay*, 26,981, 1907. *Roettger*, 7,858, 1910. *Schneider*, 2,496, 1910; 11,461, 1910. *Schmidt*, 13,304, 1909; 6,515, 1909. *Siemens*, 2,064, 1857. *Strong*, 9,142, 1900. *Sueur*, 10,722, 1900; 4,828, 1901. *Thrupp*, 26,767, 1898; 1,913, 1900. *Tofehrn*, 21,085, 1898. *Tripler*, 15,235, 1899. *Wiart*, 26,052, 1907.

German.—*Blau*, 223,843. *Chemische Fabrik Greisheim-Elektron*, 240,876. *Claude*, also *Soc. L'air Liquide*, 173,276, 177,519, 179,950, 192,594, 193,008, 202,778, 234,308, 235,422, 237,438. *Drägerwerk*, 179,793. *Flüssige Luft Heylandt*, 191,659. *Hecker*, 204,807. *Hildebrandt*, 169,331, 179,132, 181,115, 183,410, 193,007, 197,071, 198,503, 214,264, 223,064, 228,487, 229,345, 247,581, 249,996.

Jänecke, 220,270, 246,341. *Lachmann*, 167,931. *Levy*, 191,916. *Levi Cases*, 243,031. *Levi and Hellbronner*, 158,838. *Linde*, also *Ges. für Linde Eismachinen*, 88,824, 173,260, 180,014, 203,814. *Metz*, 119,943. *Merves*, 174,362, 179,782, 195,754, 199,035, 209,074, 209,848, 222,840, 227,100, 229,276, 230,002, 233,897, 238,690, 242,867, 249,997, 251,584. *Mix*, 124,376, 142,918. *Pictet*, 162,323, 162,702, 165,268, 159,934, 164,382, 169,359, 178,246, 169,564, 200,053. *Solvay*, 39,280. *Thrupp*, 115,421, 142,935. *Tripler*, 135,728.

French.—*Aumont*, 427,266. *Bobrick*, 399,416. *Claude*, also *Soc. L'air Liquide*, 296,211, 299,051, 328,245, 352,856, 322,107, 324,097, 370,850, 381,790, 338,842, 416,556, 401,496, 410,967, 322,702, 324,460, 329,839, 338,964, 423,224, 376,270, 381,866. *Dumars*, 332,662. *Hellbronner*, 376,154. *Hildebrandt*, 355,206, 373,894, 384,139, 403,630, 420,903, 417,537. *Jänecke*, 407,852. *Joly*, 318,153. *Levy*, 328,770, 330,258. *Levi Cases*, 433,639. *Levi and Hellbronner*, 328,984. *Linde*, also *Gesell. für Linde Eismachine*, 312,811, 322,811, 333,205. *Merves*, 360,427, 358,236, 425,546, 426,908. *Pictet*, 322,600, 295,002, 324,715, 409,787. *Place*, 321,860, 379,605, 391,326. *Roettger*, 413,958. *Schneider*, 416,122. *Thrupp*, 307,941.

Andrews, in 1868, found that each gas has a "critical temperature" above which it was unliquefiable whatever the pressure. It is only necessary, in order to liquefy the "permanent gases," to find this "critical temperature" and go below it.

The commercial realisation of this was due to Carl Linde, who in 1895 inaugurated a system which to-day shares with the Claude system practically the monopoly of the liquid air industry.

Theoretical Considerations.—The production of extreme cold in modern processes is obtained by the expenditure of mechanical work, namely by compressing a gas or vapour, and then allowing it to expand in such a manner that owing to the cooling produced it will absorb heat from the substance to be cooled.

When a gas is allowed to expand from a volume v_2 to a volume v_1 *against an external pressure*, the gas performs external work—

$$\int_{v_2}^{v_1} p\,dv,$$

which results in the abstraction of an equivalent amount of heat from the gas or its surroundings.

There should, however, be no heating or cooling of a *perfect* gas on *free expansion*, *i.e.*, when *no external work is performed.* Kelvin and Joule, however, found that air, oxygen, nitrogen, and carbon dioxide were all cooled on free expansion, the cooling being most noticeable in the case of carbon dioxide, the least perfect of these gases. This cooling is explained by the fact that internal work is performed during the separation of the molecules, which occurs during the expansion of the gas. In the case of hydrogen a small heating effect was noticed, but it is now known that hydrogen, on expansion, is cooled like any other gas, provided the temperature is below a certain value. The fall in temperature in degrees Centigrade in the case of air on free expansion was given by Thomson and Joule as :—

$$0.276\left(p_2 - p_1\right)\left(\frac{273}{T}\right)^2,$$

where T is the initial absolute temperature of the gas, and p_2 and p_1 the pressures before and after expansion. Thus the fall in temperature of air expanding at ordinary temperature (15° C.) from 100 atmospheres to atmospheric pressure is about 25° C.

Emil Vogel has recently shown[1] that as the pressure increases, the cooling per atmosphere drop decreases, so that at ordinary temperatures the limit is reached if the pressure is raised to about 300 atmospheres. The formula expressing the cooling effect is given as

$$dt = (0.268 - 0.00086p)\left(\frac{273}{T}\right)^2 dp.$$

The Principle of the Temperature Exchanger.—The fall of temperature of 25° C., due to a fall in pressure of 100 atmospheres (see above), is of course entirely inadequate for the spontaneous production of liquid air, the boiling point of which at atmospheric pressure is −193° C. In modern

[1] "Uber die Temperaturveränderung von Luft und Sauerstoff." Berlin, 1910.

practice, therefore, the principle of the *temperature interchanger* is applied. The essence of this principle, first suggested by Wilhelm Siemens in 1857, is as follows:—

The compressed air passes through the tube A (Fig. 1) into the expansion apparatus B, where it is consequently cooled. The expanded and cold air passes out by means of the tube C concentric

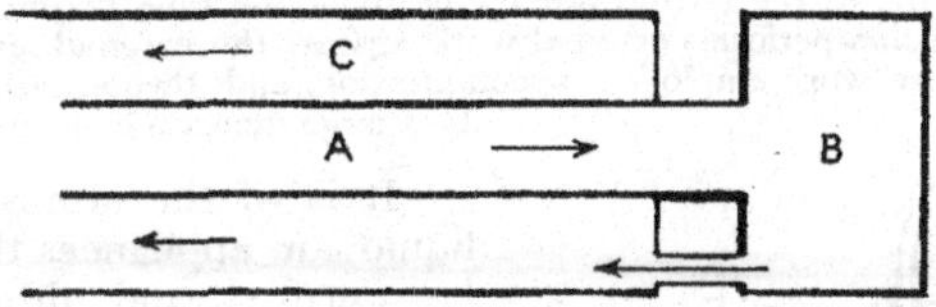

FIG. 1.—Temperature Interchanger.

with A, and thus cools the ingoing compressed air. A cumulative fall in temperature therefore results, which theoretically is only limited by the actual production of liquid air. By this means it is ensured that when the expanded air escapes into the atmosphere it is only slightly inferior in temperature to that of the compressed air entering the exchanger. In good modern exchangers the difference in temperature between the ingoing and escaping air is only 3° or 4° C.

Types of Air-Liquefying Machines.—Liquid air appliances are divided broadly into two groups—(1) those, such as the Linde and Hampson systems, which depend on *internal work*, *i.e.*, on the Joule-Thomson effect, and (2) such as the Claude process, which is a solution of the problem of expansion with *external work*.

1. The principle of machines depending on *internal work* is as follows:—

The air, compressed to a high pressure and freed from moisture and carbonic dioxide by circulating through purifying tubes, is allowed to pass by means of a tube A (Fig. 2) (in reality in the form of a long thin spiral) to the valve B, where expansion by simple outflow takes place. Cooling takes place according to the Joule Thomson effect, and the cooled air is led back through a tube C concentric with A, and finally is discharged into the atmosphere or back to the compressor. Owing to the exchange of heat from the compressed to the expanded air, the temperature of expansion

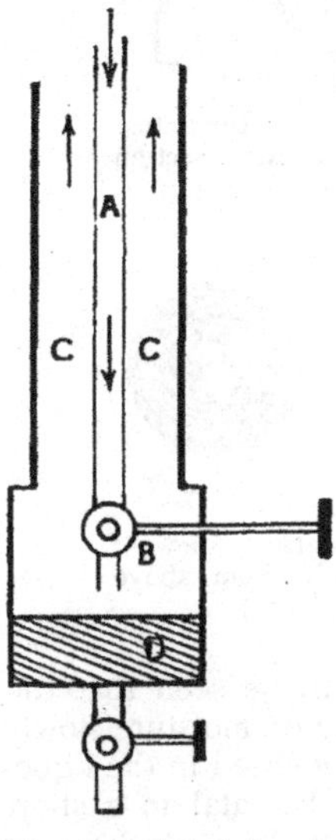

FIG. 2.—Principle of Machines depending on Internal Work.

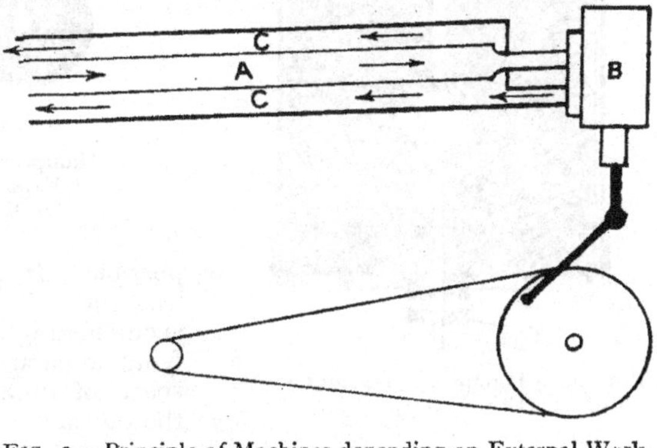

FIG. 3.—Principle of Machines depending on External Work.

is rapidly lowered, until finally the temperature of liquid air at the pressure to which the air expands is reached. From this moment some air will liquefy at the expansion valve and will collect in vessel D.

2. The second method of liquefaction, *i.e.*, that depending on *external work*, is an application of the suggestion made by Lord Rayleigh in 1898, that if the simple expansion described above were effected against the vanes of a turbine, by

thus utilising external work done by the expanding gas an increase of efficiency would result.

In actual practice the compressed air is delivered through a pipe A (Fig. 3) to the expansion machine B, which in this case is a compressed air motor working against an external resistance. The expended air returns to the compressor by means of the tube C concentric with A. By this means the gas on expansion performs external work against the external resistance, thus regaining a large proportion of the work employed in compression, and, theoretically at least, there should be a large increase in efficiency.

It is of the utmost importance in all liquid air appliances that the air supplied be purified and dried as perfectly as possible. It can easily be seen that the least trace of impurity or moisture slowly accumulating and condensed in the liquefying apparatus will be fatal in a short space of time by causing a blockage of the tubes.

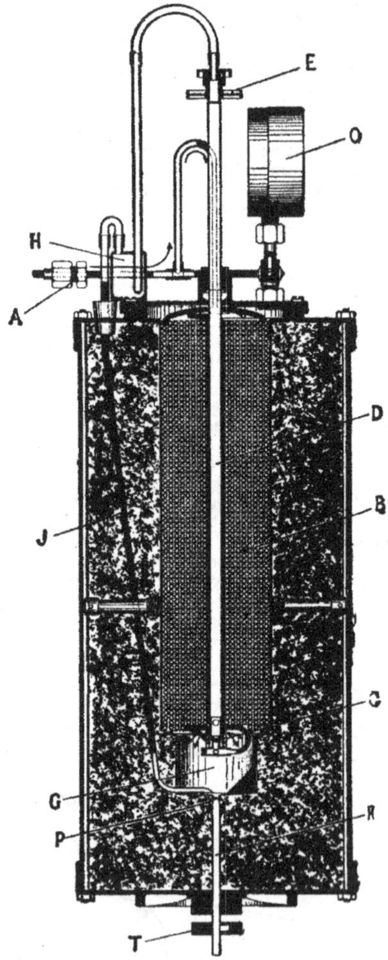

FIG. 4.—Hampson's Apparatus. Section (British Oxygen Co.).

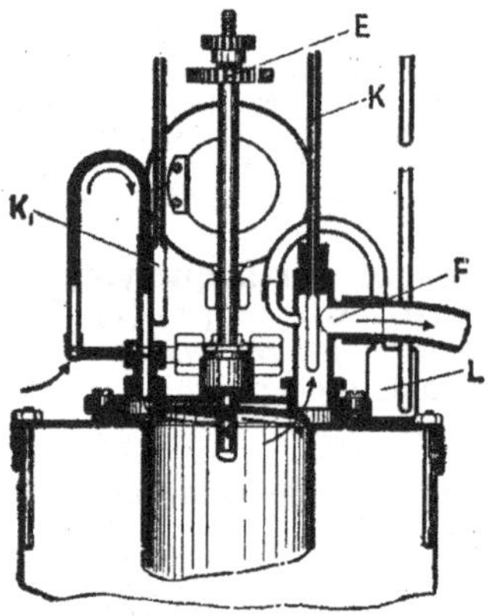

FIG. 5.—Hampson's Apparatus. Details of Upper Part. Section at Right Angles to Fig. 4.

FIG. 6. FIG. 7.

Hampson's Apparatus. View of Expansion Coils from above (British Oxygen Co.).

Hampson's Apparatus.—This apparatus, although only a laboratory appliance and of low efficiency, has the merit of simplicity and handiness. It begins to liquefy air after ten minutes' working, and produces about a litre of liquid air per hour with an expenditure of 6 H.P. on the compressor.

Figs. 4, 5, 6, 7 give sectional views of this liquefier. Air is drawn into a compressor through a purifier containing slaked lime, which removes carbon dioxide. The compressed air at a pressure of about 200 atmospheres then passes through a vessel in which most of the water picked up in the compressor is separated out, and afterwards through caustic potash, which removes the last traces of moisture and carbon dioxide.

The purified compressed air passes through the inlet A into the regenerator coils B. It travels down the coils, and escapes at the valve C, which is regulated by a hollow spindle D, provided with a hand wheel E. The air expanded down to atmospheric pressure passes back over the regenerator coils, and escapes by the pipe F (Fig. 5) back to the compressor. After about ten minutes' working, liquid air collects in the receiver G, and can be run off through the tube R by opening the valve P by means of the handle T.

The outlet air pressure is indicated by the glycerine gauge L, while a pressure gauge O registers the pressure of the compressed air. K and K_1 are thermometers indicating the temperatures of outgoing and incoming air.

The liquid receiver holds about 100 c.c. The level of the liquid in it is shown by the small gauge H, which communicates with the receiver by means of a pipe J and the hollow spindle D.

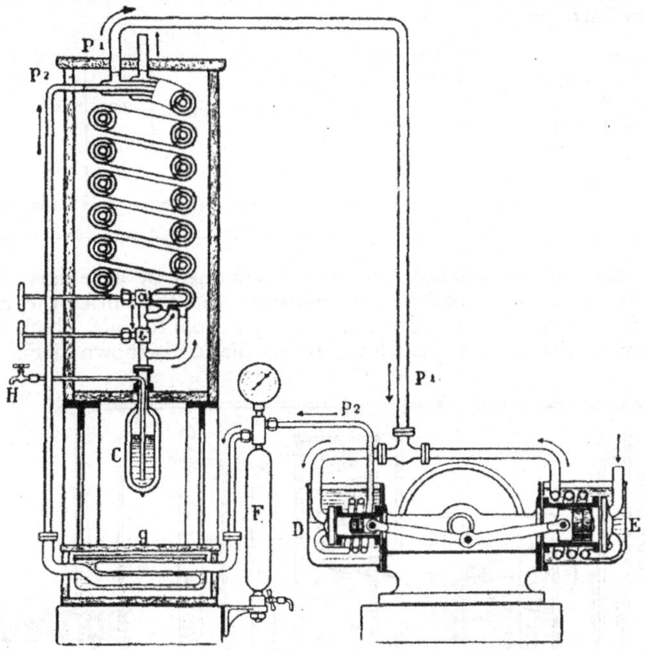

FIG. 8.—Linde's Apparatus for Liquefying Air. Laboratory Type.

When liquid air collects in the receiver G, it compresses the air in the pipe J, and so displaces the coloured liquid in the gauge glass H.

All the cold parts are carefully protected against the penetration of external heat by a thick covering of insulating material.

The advantage of this apparatus, which is absolutely practical and used in a large number of laboratories, is its extreme simplicity, there being only one gauge to watch and one valve to control. It is sold by the British Oxygen Co.

The Linde Process.—In this process, the expansion of the compressed gas takes place by simple outflow. Fig. 8 gives a diagrammatic representation of the laboratory type of Linde machine.

The compressor takes air from the atmosphere, and first compresses it to 40 atmospheres in E, and further to 200 atmospheres in D. The compressed air passes first into the iron bottle F, where most of the moisture and impurities are deposited, then through a coil *g*, placed in a refrigerating mixture, where it is cooled and deprived of the rest of its moisture by solidification. The dry compressed air then passes through the pipe P_2 into the inner tube of the temperature exchanger, which is in the form of three concentric spiral tubes. Expansion down to 40 atmospheres takes place at the needle valve *a*, which is regulated by a cock. The expanded and thereby cooled air returns by means of the middle tube of the temperature exchanger and the pipe P_1 to the compressor, where it is again compressed to 200 atmospheres. Progressive cooling is thus produced until a sufficiently low temperature is attained to liquefy some of the air expanding through the valve *a*. This liquid is allowed to pass at intervals through the valve *b* into the vessel C, where the pressure is atmospheric. This fall in pressure of the liquid produces a considerable evaporation, and therefore the evaporated gases are caused to pass through the outer tube of the exchanger into the atmosphere, thus ensuring that on their exit into the atmosphere all their cold is given up to the incoming compressed air. The amount of liquid air in the container C, which is in the form of a double-walled vacuum vessel, gradually increases, and can be run off through a tap H. The whole of the exchanger is covered by a wooden casing, filled with sheep's wool lagging, to prevent the entrance of heat from the surroundings.

A machine of this type producing 50 litres per hour requires 2 H.P. per litre, and begins to produce liquid after about ninety minutes' working. The following table indicates the increase in efficiency produced by the initial cooling of the compressed gas:—

Output, litres per hour	With previous cooling	0.5	1.25	2	3	5	10	20	50	100
	Without previous cooling	...	...	...	...	...	5.5	12.5	35	70
Power used, H.P.		3.5	5.5	8.5	12	19	30	52	105	190
Cooling water used, litres per hour		250	400	600	850	1,400	2,300	3,800	8,000	15,000

This cooling in the smaller machines is obtained by a refrigerating bath as shown above, which also solidifies the moisture and thus dries the air. In the larger machines an ammonia compressor is used to cool the air, which is first dried by passing over calcium chloride. Such an installation is shown in Fig. 9.

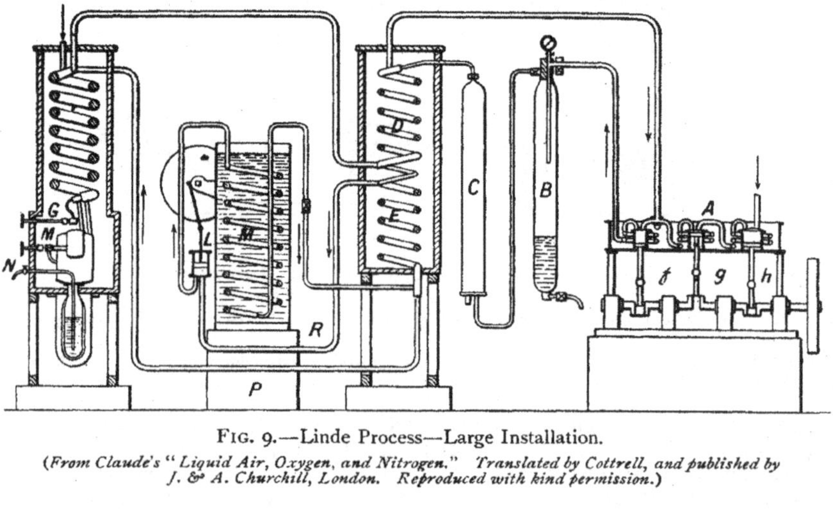

FIG. 9.—Linde Process—Large Installation.

(*From Claude's "Liquid Air, Oxygen, and Nitrogen." Translated by Cottrell, and published by J. & A. Churchill, London. Reproduced with kind permission.*)

For large installations a three stage compressor is used. The air compressed to 200 atmospheres in the high pressure cylinder *f* loses most of its moisture in B and is completely dried over calcium chloride in C. It then passes through the inner tube of the cooler DE by means of the pipe to the inner tube of the exchanger F which has been described above, p. 7. The air at about 50 atmospheres coming from the middle tube of the exchanger F is passed through the outer tube in the top half of the spiral cooler D, thus producing a preliminary cooling of the compressed air, and so to the compressor where it is again compressed to 200 atmospheres. The auxiliary cooling is effected by an ammonia compressor L. The compressed ammonia is cooled and liquefied in the water bath M. The liquid ammonia expands through the outer tube of the cooler E, thus cooling the compressed air passing to the exchanger F. The expanded gaseous ammonia is then returned to the ammonia compressor through the pipe R.

Carbon dioxide is removed from the air by passage over lime or caustic soda.

Recently Professor Linde has used another method of desiccating the air by means of cold. The rime is condensed in the exchangers which are changed over every twenty-four hours, thus allowing the apparatus to work continuously for twelve days or more.

Theory of the Linde Process.—In the case of the expansion of a perfect gas by simple outflow without sensible velocity from a pressure p_1 to a pressure p_2, the work done is always against the pressure p_2, and thus the work involved in the expansion is—

$$\int pdv = p_2\int dv = p_2v_2,$$

where v_2 is the expanded volume.

From a consideration of Boyle's Law it can be shown that this work is not translated into cooling.

Air, however, is far from being a perfect gas, and instead of assuming that the relation of temperature, pressure, and volume is given by Boyle's Law, viz., $pv=RT$, where P_1 is a constant, the facts are far more closely expressed by Van der Waals' equation—

$$\left(p + \frac{a}{v_2}\right)(v - b) = RT,$$

where a, b, and R are constants.

If, therefore, an imperfect gas expands from a volume v_1 to a volume v_2, the work done due to internal pressure of the molecules is—

$$\int_{v_1}^{v_2} pdv = \int_{v_1}^{v_2}\frac{a}{v^2}dv = \frac{a}{v_1} - \frac{a}{v_2}.$$

If, as is the case, practically the volume v_2 is great compared to v_1, the "internal work" due to expansion is $\frac{a}{v_1}$.

A full consideration of the problem shows that the whole practical cooling effect or expansion by simple outflow is due to this external work.

This conclusion is somewhat modified in practice by the fact that the compressed gas before expansion first passes through the temperature exchanger, but a complete analysis of the facts supports the view that the whole cooling effect depends on the state of the compressed air at entry to the exchanger, the cooling effect being equivalent to $\frac{a}{v_1}$, where v_1 is the volume of the gas at the entry to the exchanger. In order to increase the yield the quantity $\frac{a}{v_1}$ should be increased as much as possible. Obviously in order that v_1 shall be small the pressure of the gas before expansion must be high. This is the case in the Linde process, the air being compressed to about 200 atmospheres. A considerable increase in efficiency is obtained by expanding down only to about 50 atmospheres instead of right down to atmospheric pressure, for as by this means the work of compression from 50 to 200 atmospheres is far smaller than is the case from 1 atmosphere to 200 atmospheres the cooling effect on expansion in the two cases is practically the same. Thus the work of compression from a pressure of 1 atmosphere to 200 atmospheres is proportional to $\log\frac{200}{1}=2.3$, whereas the work of compression from 50 atmospheres to 200 is proportional to $\log\frac{200}{50}=0.6$ only.

Linde, therefore, expands from 200 to 50 atmospheres, the expanded gas being returned to the compressor at this pressure.

As the cooling effect depends on the state of the gas at entry to the exchanger, an increase in efficiency is also obtained by cooling the compressed gas before it enters the exchanger, thus decreasing v_1 and increasing the value $\frac{a}{v_1}$. The Linde process actually employs an auxiliary cooling machine which cools the gas after compression, and before passing into the exchanger.

The Claude Process.—In 1902, after several years of experiment, Claude first liquefied air by allowing compressed air to expand in an ordinary compressed air motor, thus producing extreme cooling effects due to the external work done by the expanding gas on the piston of the motor.

One of Claude's greatest difficulties during his work was the lubrication of the moving parts of the machine when at the low temperatures attained. This difficulty was at first obviated by the use of petroleum ether, which at these low temperatures acts as an excellent lubricant. When the machine was started up, ordinary lubricating oil was used, this being gradually mixed with increasing quantities of petroleum ether as the temperature fell. In 1912, however, Claude succeeded in abolishing all lubrication by the use of suitably prepared leather stampings on the working parts. This leather was found to retain its ordinary properties at these low temperatures, and required no lubrication whatever.

Claude's first attempt met with ill-success, owing to the fact that, as air becomes more and more imperfect as the temperature decreases, he was obtaining only a slight expenditure of external work due to the expansion of the cooled gas. This imperfection is also enhanced by the increase of specific heat of air with decrease of temperature, which limits the fall of temperature due to the expansion.

The solution of this difficulty was found in slightly raising the initial temperature of expansion. Claude's method of obtaining this result is shown in Fig. 10.

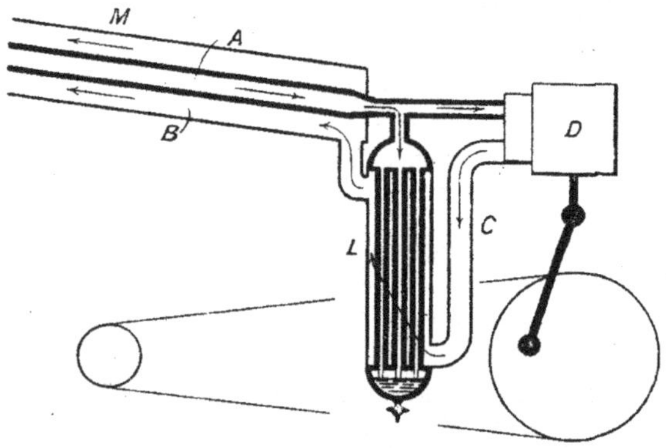

FIG. 10.—Claude Process.

(*Reproduced from Claude's "Liquid Air, Oxygen, and Nitrogen," with kind permission of the publishers, Messrs J. & A. Churchill.*)

The compressed air at a pressure of 40 atmospheres passes through the inner tube A of the exchanger M to the expansion machine D. The expanded and cooled air then passes upwards round the outside of the tubes of the liquefier L. These tubes are supplied with the compressed air at 40 atmospheres from the tube A. This compressed air is thus progressively cooled by the expanded gases circulating upwards until the temperature of liquefaction at 40 atmospheres, viz., about $-140°$ C., is attained. Liquefaction then commences in the tubes, the liquid collecting in the bottom of the liquefier from which it can be run off by means of a cock. The expanded gas passes round the tubes of the liquefier and thus attains the temperature of liquefaction of the compressed gas; it then passes into the outer tube B of the exchanger, and thus cools the incoming compressed gas, which, therefore, reaches the expanding machine at this temperature. In this way it is ensured that the initial temperature of expansion does not fall so low that the gas becomes too imperfect to produce liquefaction. In order that this temperature should remain at the desired height the liquid formed in the liquefier must be run off at intervals.

In more recent machines (see Fig. 11) the advantages gained by the above procedure are increased by the method of compound liquefaction in which the expansion takes place in two stages. The compressed air passes through the tubes of the exchanger into the first expansion machine. The partially expanded gas then circulates round the top part A of the liquefier tubes which are fed with compressed gas by means of the tube S. On leaving the top of the liquefier the partially expanded gas undergoes a second expansion in C_2, on exit from which it flows round the bottom half of the tubes of the liquefier, then by means of the outside tubes of the exchanger M back to the compressor. The liquid formed in the tubes of the liquefier collects in the vessel D. The compound expansion machine is arranged with the two expansion cylinders on the same shaft, using a single piston, thus obviating difficulties due to leaks on the high pressure cylinder.

TABLE OF OUTPUT OF CLAUDE MACHINES.

Output. Litres per hour.	H.P. used.
5	15
50	60
65	75

The advantages of the Claude process are :—

1. The use of comparatively low pressures of the compressed gas.
2. Short time necessary to attain the temperature of liquefaction.

Consequently the process is increasingly in favour.

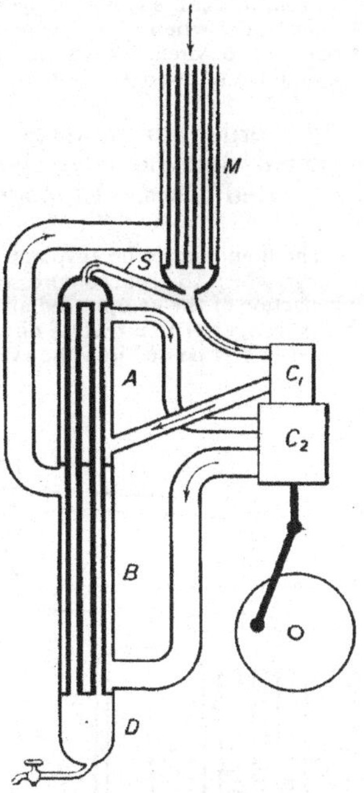

FIG. 11.—Claude Process, with Expansion in Two Stages.

(Reproduced from Claude's "Liquid Air, Oxygen, and Nitrogen." by kind permission of the publishers, Messrs J. & A. Churchill.)

The Separation of the Constituents of Liquid Air.—The basis of the separation of the elements of air by the method of liquefaction was first suggested by Parkinson in 1892 and depends on the difference in their volatility, the boiling points under atmospheric pressure being − 182.5° C. for oxygen and − 195.5° C. for nitrogen. The problem, therefore, is not very different from the separation of a mixture of alcohol and water, although the simple fractional evaporation of liquid air gives very poor yields, and it was not until Linde in 1902 utilised the method of rectification that the process became one of commercial importance.

The following table of Linde's experimental results gives the composition of the liquid and gaseous states as liquid air slowly evaporates.

Per Cent. Liquid Evaporated.	Per Cent. Oxygen in Liquid Phase.	Per Cent. Oxygen in Gaseous Phase.	Per Cent. Oxygen Left in Liquid.
0	23.1	7.5	100
50	37.5	15.0	80
70	50.0	23.0	65
80	60.0	34.0	52
85	67.5	42.0	43
90	77.0	52.0	33
95	88.0	70.0	19

From this it will be seen that in ordinary evaporation to obtain an oxygen residue of 50 per cent. purity, 70 per cent. of the liquid must be evaporated with a loss of 35 per cent. of the total available oxygen.

Similar results were obtained by Baly in 1900.

Temperature.	Per Cent. Oxygen in Liquid.	Per Cent. Oxygen in Vapour.	Temperature.	Per Cent. Oxygen in Liquid.	Per Cent. Oxygen in Vapour.
− 195.46	0	0	− 188	72.27	44.25
− 195	8.10	2.18	− 187	77.80	52.19
− 194	21.6	6.8	− 186	82.95	60.53
− 193	33.35	12.0	− 185	87.60	69.58
− 192	43.38	17.66	− 184	91.98	79.45
− 191	52.17	23.60	− 183	96.15	89.80
− 190	59.55	29.93	− 182	100	100
− 189	66.20	36.80			

From this table it will be seen that the vapour phase is richer in nitrogen than the liquid phase. For instance, when liquid air (21 per cent. O, 79 per cent. N) evaporates, the vapour contains some 7 per cent. oxygen. Thus in simple rectification, although all the nitrogen can be expelled from the liquid, the nitrogen itself will contain at least 7 per cent. oxygen.

In order to produce the evaporation of the liquid air economically, it is essential that no cold should be lost either from the liquid or from the cold evaporated gases. In modern processes this is ensured as follows:—

The liquid air to be evaporated is contained in the vessel V (Fig. 12), containing a series of tubes F through which the cold compressed air is passed. The cold from the evaporating liquid air in V causes liquefaction of the compressed air in the tubes, this air having already been cooled by passage through the exchanger H in a reverse direction to the cold vapours from the evaporating liquid. The liquid formed in C is passed into the evaporating vessel V by means of the tube R.

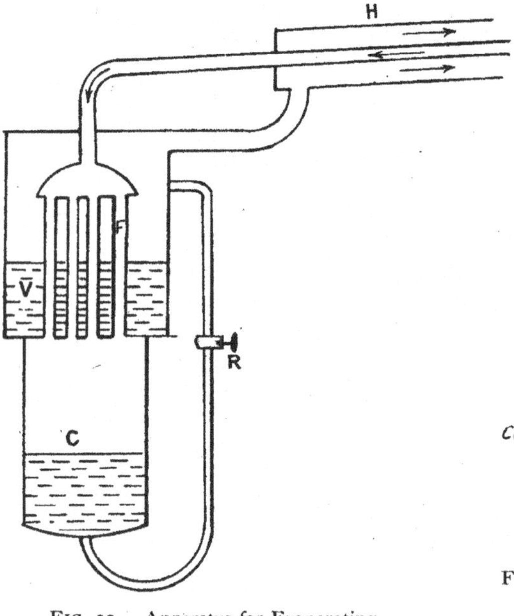

FIG. 12.—Apparatus for Evaporating Liquid Air.

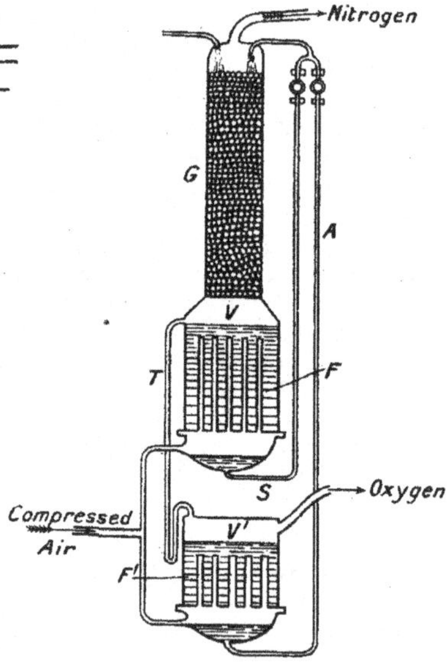

FIG. 13.—Linde's Apparatus for Oxygen.

(From Claude's "Liquid Air, Oxygen, and Nitrogen," by kind permission of the publishers, Messrs J. & A. Churchill.)

Although by this means evaporation is produced with the penetration of very little heat from the surroundings, yet the amount of liquid formed from the compressed air is necessarily not equivalent to that evaporated, and therefore it is necessary to make up the loss by the continuous addition of liquid air.

Linde's Apparatus for Pure Oxygen.—Linde's method of rectification is shown in Fig. 13. The liquid air to be evaporated is contained in V and its evaporation is produced by the liquefaction of the compressed air in the tubes F contained in V. The liquid air thus formed is discharged by the tubes A at the top of the rectifying column up which the vapour from the evaporating liquid in V ascends. The descending liquid, containing about 21 per cent. oxygen, loses its nitrogen easier than its oxygen and thus becomes colder. It therefore condenses the oxygen in the ascending gases and thus gets richer in oxygen as it descends. When the apparatus is working normally, therefore, the liquid in V is practically pure oxygen. When this is the case there will be a temperature gradient established in the column, the maximum cold being at the top. The liquid in V can overflow continuously into a second evaporator V^1 from which it evaporates through a pipe S with a purity of about 98 per cent., the cold produced by the evaporation causing liquefaction of compressed air in the tubes F^1, this liquid being also passed to the top of the rectifying column. The nitrogen, which, as already explained, must contain at least 7 per cent. oxygen, escapes at the top of the column.

A Linde separator capable of producing large quantities of oxygen of about 99 per cent. purity is shown in Fig. 14.

Air at 2,000 lbs. square inch pressure passes by means of the pipe D through the three small tubes of the temperature exchanger and the spiral tube d_1 in the vaporiser to the expansion valve G. The air expands here and is discharged by means of the pipe d_2 at the top of the rectifying column. The expanded air then passes by means of the tubes c_1 and e_1 through the pipes C *c* E of the temperature exchanger in the reverse direction to the incoming compressed air. As soon as the temperature at the expansion valve falls to the liquefaction temperature, liquid air is produced which collects in the vaporiser B. As this accumulates it evaporates, due to the heat from the compressed air in the tube d_1, and the vapour ascends against the downward flow of liquid air from the pipe d_2. Rectification takes place until finally the liquid in B is practically pure oxygen. The excess of this oxygen is continuously drawn off by means of the pipe e_1, through the tube *e* of the exchanger to the exit E. The nitrogen passes by means of F through the tubes c_1 and *c* to the exit C.

The whole of the apparatus is cased in with wood, all spaces being filled with sheep's wool to prevent penetration of heat from the surroundings.

The nitrogen still contains 7 per cent. oxygen, and therefore in this system at best only 74 per cent. of the available oxygen is obtained.

A complete oxygen plant on the Linde system is shown diagrammatically in Fig. 15.

The air enters the purifier B at A. Here it is decarbonised by means of milk of lime. The purified air compressed in the compressors D, and then cooled in water, passes through the pipe E to the driers F. Here drying is effected first by chloride of lime and finally by potash. The dried air is then cooled in G by means of the refrigerating machine H, and passed into the separating apparatus K. As the desiccation of the air is never complete, it is necessary for continuous working to duplicate the separator K. The oxygen escapes by means of the tube L.

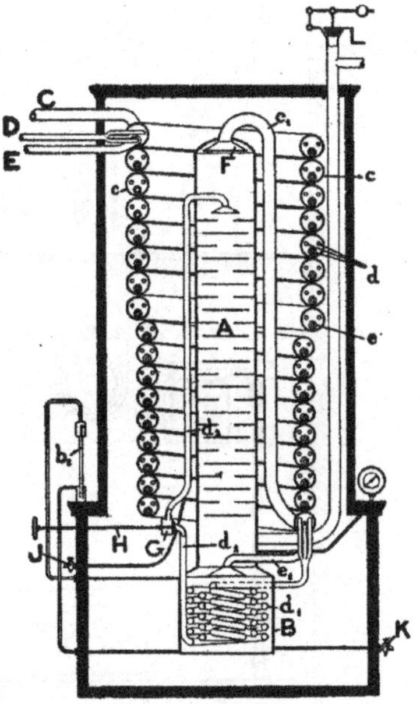

FIG. 14.—Linde Separator.

(*Taken from Thorpe's "Dictionary of Applied Chemistry," by kind permission of the publishers, Messrs Longmans & Co.*)

Of late Linde has desiccated the air by means of the auxiliary refrigerating machine. In this case the exchangers in which the moisture is deposited as rime are duplicated and changed over about every twenty-four hours.

The various sizes of oxygen machines working on the Linde system are given in the following table:—

	I.	II.	III.	IV.	V.	VI.	VII.	VIII.	IX.	X.
Output of oxygen, cubic metres per hour	1	2	5	10	20	50	100	200	500	1,000
Power used, H.P. - - - - - -	6	10	20	35	55	100	185	350	800	1,500
Cooling water used, cubic metres per hour	0.25	0.5	1.0	1.6	2.4	4.0	7.0	13.0	29.0	54.0

Production of Nitrogen by the Linde System.—The nitrogen obtained from the foregoing apparatus contains at least 7 per cent. oxygen. In order to produce pure nitrogen Linde has modified the method of rectification. In the earlier apparatus the liquid produced in the worm d_1 (Fig. 14), instead of being discharged at the top of the column, is discharged at a point half-way down. Thus in the top half of the column the gas consists of nitrogen containing some 7 per cent. oxygen. The liquid oxygen in the vaporiser is drawn by suction through a worm situated in this nitrogen. The cold produced by the evaporation of this oxygen causes part of the vapours ascending the reclining column to condense, thus reducing the amount of oxygen, and producing practically pure nitrogen at the top.

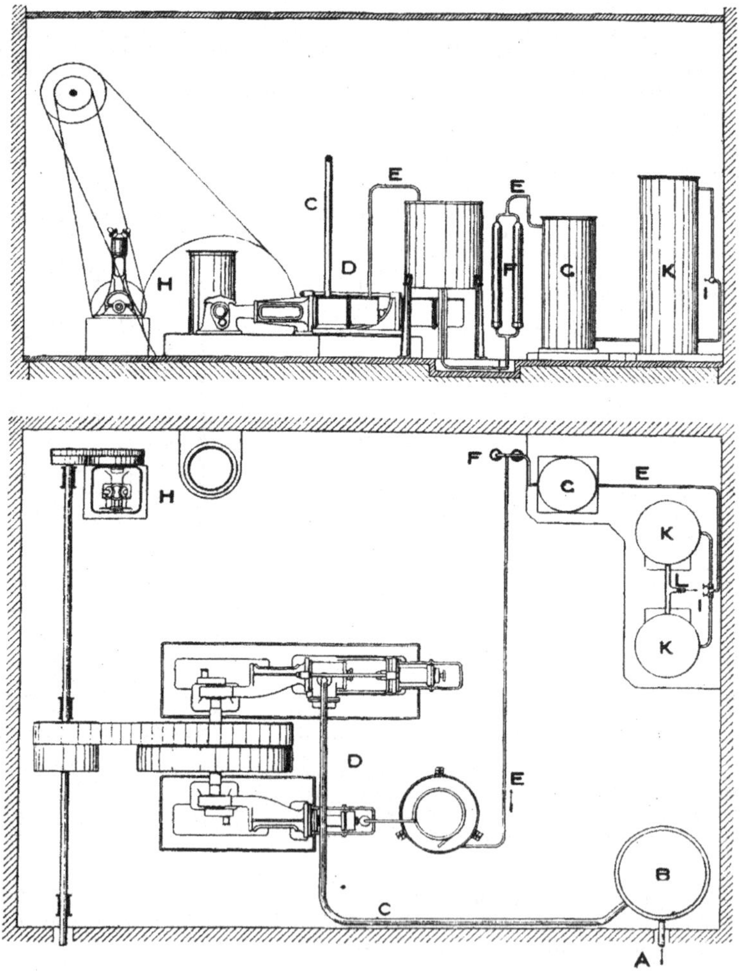

FIG. 15.—Linde Complete Oxygen Plant.

Although this method allows the waste oxygen to carry away some 40 per cent. nitrogen, it is still used owing to its simplicity in smaller plants.

A more economical method is shown in Fig. 16.

The apparatus consists of the exchanger C, the rectifying column A, and the vaporiser B. The whole apparatus is cased in with wood, and filled with sheep's wool to prevent the entrance of external heat.

The temperature exchanger, which is in the form of a spiral, consists of three small pipes, *e d f*, encased in a large tube C. The bottom end of the external tube C is carried up to the top of the rectifying column.

The cooled and purified air at a pressure of about 2,000 lbs. per square inch is passed by means of the tube *f* through the exchanger, then through the coil f_1 in the vaporiser, to the expansion valve H. The air here expands, and passes into the rectifying column. It flows out at the top of the rectifying column back through the temperature exchanger by means of the pipe C. On exit, part of this air is led back to the compressor, and is again compressed. The usual progressive cooling effect is thus produced, and eventually liquid air is produced, and collects in the vaporiser B. Owing to the process of rectification in the column A the liquid in B will become nearly pure oxygen, and the nitrogen escaping at the top of the column, owing to the fact that it is traversing a closed circuit, will tend to become purer and purer. The additional supply of air necessary for the continuous working of the apparatus is supplied by a subsidiary compressor working at 60 lbs. per square inch. This cooled and purified air passes through the exchanger by means of the pipe *d*, is liquefied in the coil d_1 situated in the vaporiser, and is discharged, by means of the pipe d_2, into the lower part of the rectifying column. This falling liquid will become rectified by the ascending gases from the vaporiser to a purity of 7 per cent. oxygen, the descending liquid becoming nearly pure oxygen. The nitrogen containing 7 per cent. oxygen passes in part, as already described, back to the compressor by means of the pipe C, is compressed, liquefied, and discharged at the top of the rectifying column. A reference to Baly's figures will show that this liquid, containing 7 per cent. oxygen, will purify the ascending gases to 2 per cent. oxygen. These purified gases again traverse the high compression circuit, and so on. Thus pure nitrogen will eventually pass from the exit of the tube C. The oxygen can be obtained from the tube *e*.

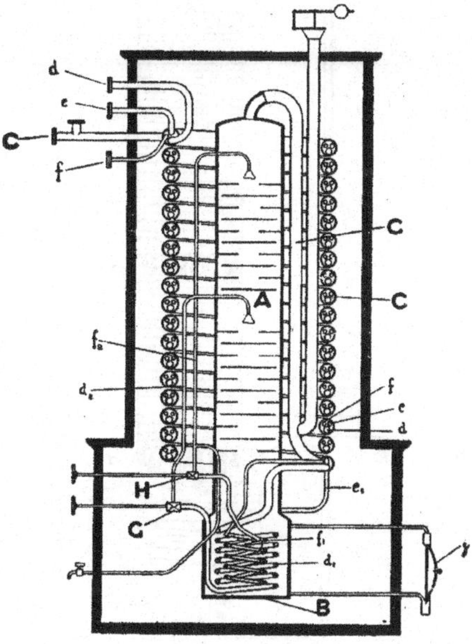

FIG. 16.—Linde Nitrogen Separator.

(From Thorpe's "Dictionary of Applied Chemistry." By the courtesy of Messrs Longmans & Co.)

Sizes of the Linde nitrogen machines are as follows :—

Size of Machine.	I.	II.	III.	IV.	V.	VI.	VII.	VIII.	IX.	X.
Output, cubic metres nitrogen per hour	6	12	30	60	120	300	600	1,200	3,000	6,000
Power used, H.P. - - - - -	9	15	30	52	80	150	275	500	1,100	2,000
Cooling water used - - - -	0.75	0.9	1.5	2.6	4	7.5	12	20	38	75

The amount of liquid used in the closed circuit can be made as large as is desired, and in consequence the purity of the nitrogen can be raised from 99.8-99.9 per cent. An apparatus working on this system has the advantage over the apparatus described previously in that it yields a gas of more constant purity. These plants take a comparatively large amount of power (about 0.5 K.W. per cubic metre of nitrogen), as the energy required for compressing the nitrogen is only partially utilised for the separation of the components.

The largest nitrogen plant which is run on the Linde system is at Odda. This plant, erected in 1908, is run by a 200 H.P. electric motor, and produces 3,000 cub. ft. of nitrogen per hour.

New Linde Apparatus for the Production of Pure Oxygen and Nitrogen.—In the oxygen apparatus described above the nitrogen leaves the apparatus with at least 7 per cent. of oxygen, so that only about 74 per cent. of the total oxygen in the air can be extracted. In order to obtain a better yield in this respect another type of apparatus has been built which subjects the air to a pre-rectification under a pressure of 4 atmospheres.

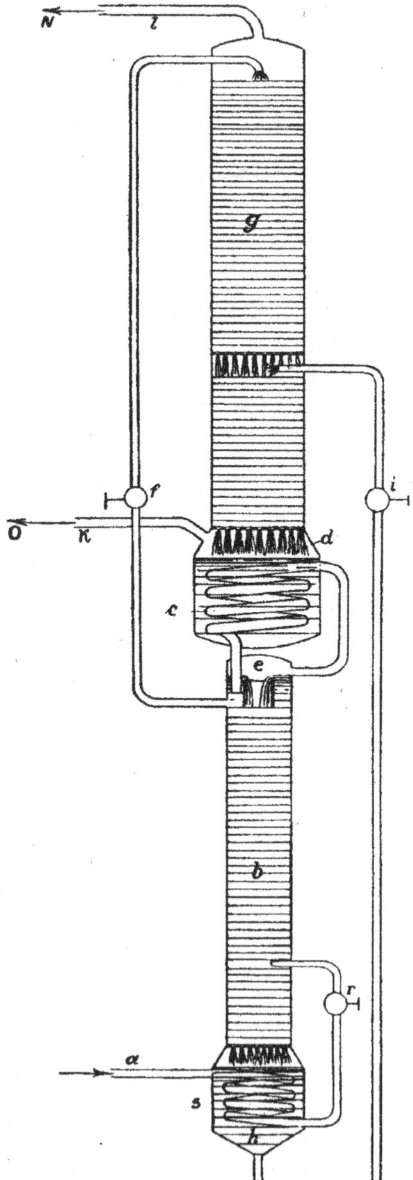

FIG. 17.—New Linde Apparatus for Oxygen and Nitrogen.

This apparatus consists of two columns, one above the other, the lower of which works at 4 atmospheres' pressure, the upper at atmospheric pressure. The compressed air entering at *a* is liquefied in the evaporator *s*, and is discharged about half-way up the lower column. The vapours rising from the evaporator *s* are purified to about 10 per cent. oxygen content by this falling liquid (the conditions of equilibrium between liquid and vapour under higher pressure being different from those at atmospheric pressure), and are further purified by the liquid which is produced in the coil situated in the evaporator *c*, so that eventually nearly pure liquid nitrogen is produced in the coil in *c*. About half of this liquid trickles back into the lower column, the other part being expanded through the valve *f* and discharged at the top of the low pressure column *g*. About half-way down this column the liquid in *s*, containing 50-60 per cent. oxygen, is discharged by means of a valve *i*. Thus in this column the final rectification takes place, the nitrogen escaping at *l*, the oxygen at *k*.

Although theoretically it should be possible to produce simultaneously pure oxygen and pure nitrogen, only 85-90 per cent. of the total oxygen is actually obtained by this means. This arrangement is capable of producing oxygen with an expenditure of energy of 1.3 K.W.H. per cubic metre.

This two-column process has also been adapted to the production of pure nitrogen. The nitrogen with a purity of 99.6-99.8 per cent. requires an expenditure of energy of 0.4 K.W. per cubic metre.

Claude's Apparatus for the Production of Pure Oxygen and Nitrogen.—In Claude's apparatus a method is employed which is styled by the inventor as the method of the "backward return," by which means a separation into a liquid rich in oxygen and one poor in oxygen is obtained. This method is illustrated in Fig. 18. The compressed air passes through the inner tube of the exchanger M to a vessel C. It then passes up the tubes of the vaporiser V, back through the outer tubes of the exchanger. The vaporised gas contains evaporating liquid air, the vapours of which also pass through an outer tube of the exchanger. The compressed air entering the lower end of the tubes in the liquefier becomes liquefied. Now a reference to Baly's results will show that in order that evaporating liquid air should furnish a gas of 21 per cent. oxygen content it must itself be of an oxygen content of 47 per cent. Conversely, if atmospheric air of 21 per cent. oxygen content is progressively liquefied, the first drops which are formed will be of 47 per cent. oxygen content. The liquid formed in the lower part of the tubes is therefore of 47 per cent. oxygen content. Thus the remaining air rich in nitrogen ascends the tubes, and in ascending is further robbed of its oxygen, until finally at the top of the tubes the gas is practically pure nitrogen. The liquid falling downward through the tubes at the same time gives up its nitrogen in exchange for the oxygen which is condensed in C. This liquid of 47 per cent. oxygen content is continuously discharged into the vaporiser V, the gases passing through the outer tube N of the exchanger.

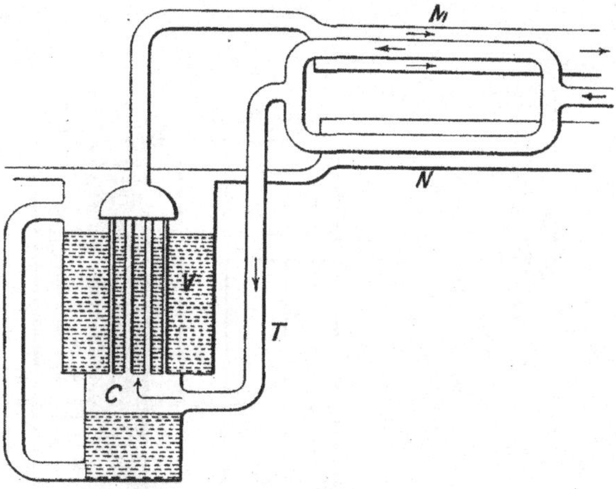

FIG. 18.—Claude's Apparatus for Pure Oxygen and Nitrogen.

(From Claude's "Liquid Air, Oxygen, and Nitrogen," translated by Cottrell. By kind permission of the publishers, Messrs J. & A. Churchill.)

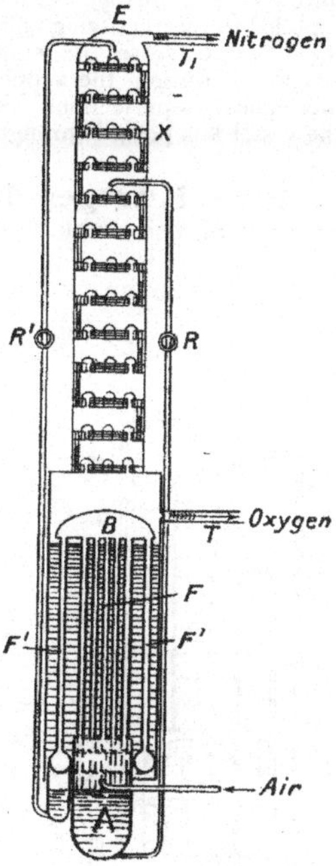

FIG. 19.—Claude's Complete Apparatus, combining the Process of Rectification with Backward Return.

(From Claude's "Liquid Air, Oxygen, and Nitrogen," translated by Cottrell. By kind permission of Messrs J. & A. Churchill.)

The complete arrangement of Claude's apparatus, combining the process of rectification with the backward return, is shown in Fig. 19. The compressed air cooled in the exchangers enters at the bottom of the series of concentric tubes F immersed in liquid oxygen and is liquefied, yielding a liquid of 47 per cent. oxygen content, which collects in A, and practically pure gaseous nitrogen. This gaseous nitrogen, passing through the concentric tubular space F^1 is, owing to its pressure, liquefied in C. This liquid nitrogen is discharged at the top of the rectifying column, while the liquid from A is discharged from the middle of the column. This liquid of 47 per cent. oxygen content will exhaust the ascending gases to 21 per cent. oxygen content. The rectification will be completed by the liquid nitrogen descending from the top of the column. Pure nitrogen will therefore pass from T_1 and pure oxygen from T. The additional liquid air necessary to compensate for the continued loss of cold is added from a separate liquid air machine.

In later machines Claude has succeeded in performing the separation of the gases and the production of the additional liquid necessary in a "single cycle."

The cooled compressed air passes through the temperature exchangers E and E[1] (Fig. 20) traversed by the cold separated gases, and divides into two portions at the point X. One portion expands in the motor O with production of external work, is liquefied and separated into liquid rich in oxygen and nitrogen in the vessels A and A[1] respectively, as explained above, these liquids being supplied to the rectifying column, as shown at different heights. Cold nitrogen, passing off through the pipe G, traverses the temperature exchanger L and liquefies the second portion of the compressed air, thus forming the supplementary liquid which is admitted to the vessel A to make up for mechanical imperfections. The nitrogen then passes through the exchanger E. The oxygen from the vessel B is passed through the exchanger E[1].

Pure Nitrogen by the Claude Process.—Although in principle the process of the backward return achieves the complete separation of the air into

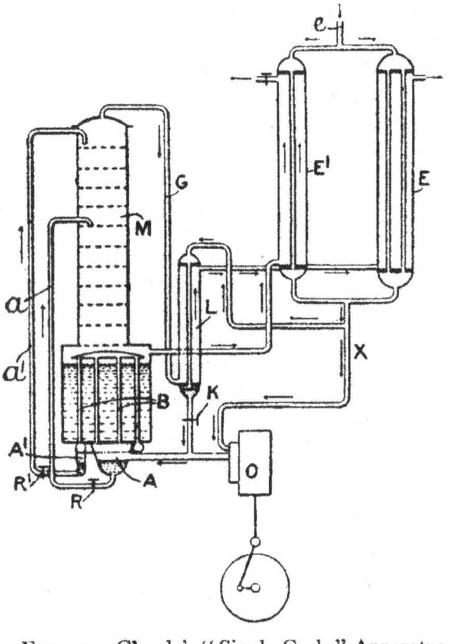

FIG. 20.—Claude's "Single-Cycle" Apparatus for producing Pure Oxygen and Nitrogen.

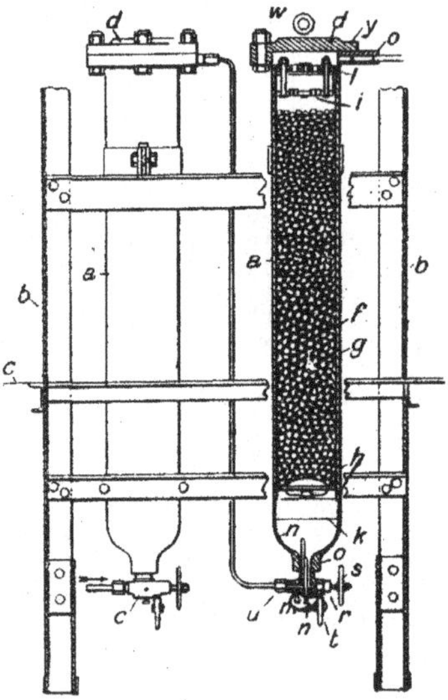

FIG. 21.—Drying Battery for Hildebrandt's Process.

pure oxygen and nitrogen, yet in practice, especially if it is desired to produce pure oxygen also, the nitrogen is only about 3 per cent. pure.

Claude has got over this difficulty by arranging that the liquefaction of the residual nitrogen at the top of the tubes of the liquefier V (Fig. 18), is facilitated by a supplementary liquefier supplied with liquid from the bottom of the rectifying column. In this way nitrogen containing not more than 0.2 per cent. of oxygen can be obtained.

In the installations working on the Claude principle the desiccation of the air is obtained by passage through soda towers, or through baffle boxes filled with hydrated lime, and finally by chloride of lime. Claude has recently succeeded in producing desiccation by cold obtained by expansion with external work in which the special exchangers are charged over every twenty-four hours as in the Linde system.

The present installations are based upon the "backward return" and double rectification, and are worked on the "single cycle" system. Recently the apparatus has been made more compact by combining the whole apparatus in a single column.

The most usual sizes in Claude's system are from 5-20 cub. m. per hour capacity, which at present are at work in some forty installations.

Hildebrandt's Process for Preparing Oxygen and Nitrogen.—This system is also coming rapidly into use, and, like Linde's process, depends on intensive cooling.

The air before entering the apparatus is freed from carbon dioxide (CO_2) and water vapour by passing over caustic soda or caustic potash (preferably the last), and then over calcium chloride in

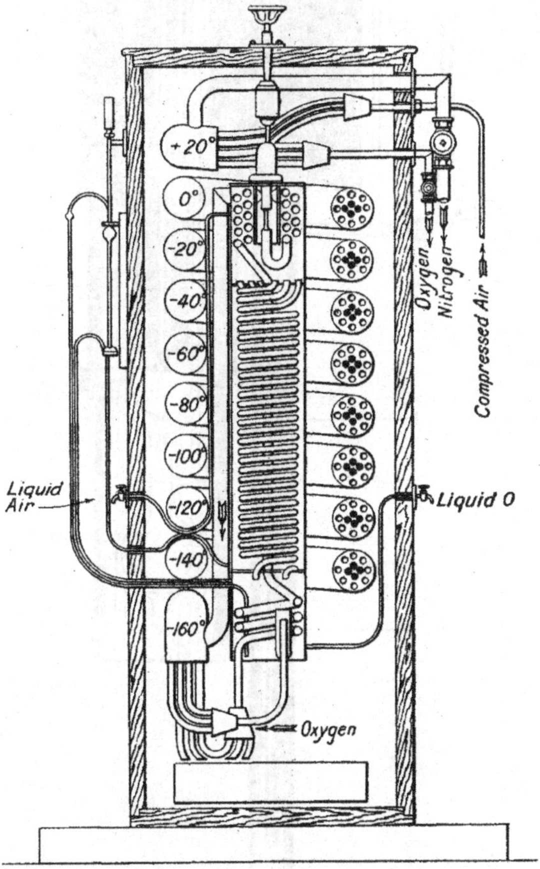

FIG. 22.—Hildebrandt's Process for Oxygen and Nitrogen.

the drying battery shown in Fig. 21. Next the air is compressed to 200 atmospheres, cooled, and then sent into the upper end of the apparatus (Fig. 22), where it is cooled by intensive self-cooling until the air liquefies.

Then by evaporation and fractional boiling it is separated into its components, oxygen and nitrogen. The nitrogen usually escapes as a by-product into the air, while the oxygen, of 98 per cent. purity, is conducted into a gasholder, and there by means of the oxygen compressor compressed into steel bottles or cylinders.

Further details should be sought in Hildebrandt's patents (see list, p. 3).

THE LIQUEFACTION OF HYDROGEN

The principle of self-intensive cooling has now been applied to the production of liquid hydrogen, and Fig. 23 shows diagrammatically such an apparatus manu-

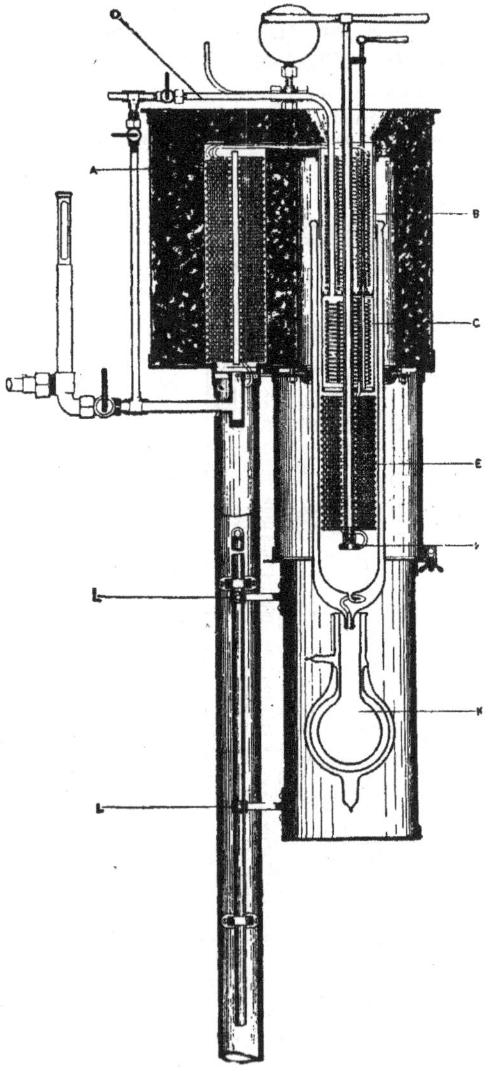

FIG. 23.—Hampson's Apparatus for Liquid Hydrogen, showing attachment of Glass Vacuum Jacketed Collecting Vessel (British Oxygen Co.).

factured by the British Oxygen Co. Hydrogen from a gasholder is compressed to about 200 atmospheres, and after purification over caustic potash enters the lower end of the coils in the chamber A. It then passes down through the coils B, which are immersed in liquid air. The compressed hydrogen, thus cooled to about − 190° C., passes through the coils C. Liquid air is allowed to drop into the chamber round these coils through a small valve regulated by a spindle at the top of the apparatus. A partial vacuum is maintained in this chamber by means of a small pump connected to the pipe O, so that the temperature of the evaporating liquid air is lowered to below − 200° C., and the compressed hydrogen is thus farther lowered in temperature. The compressed hydrogen then flows through the coil E, and expands through the valve F. The expanded gas then passes back over the coils E, round the chambers enclosing C and B, over the coils A, and back to the holder. After only a few minutes working part of the expanding hydrogen is liquefied, and can be collected in the vacuum vessel K, which can be removed at

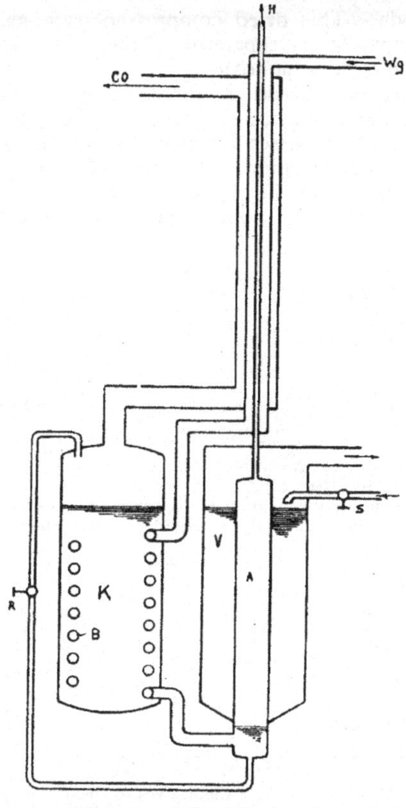

FIG. 24.—Linde-Frank-Caro Process for producing Hydrogen from Water-Gas.

will. This apparatus is capable of producing about 2 litres of liquid hydrogen per hour with an expenditure of 10 litres of liquid air.

The Linde-Frank-Caro Process for the Production of Hydrogen from Water-Gas (see also p. 39, under **Hydrogen**).—In the separation of gaseous mixtures, in which the boiling points of the constituents are very different, the liquefaction of the whole mixture is not necessary. The separation can be effected by condensing the less volatile constituent by suitable temperature and pressure, while the more volatile constituent remains in the gaseous state. A process on these lines has been applied to the separation of hydrogen from water-gas. The low temperature required is obtained by the aid of liquid nitrogen supplied from a separate apparatus. The construction of the separator is shown diagrammatically in Fig. 24.

The water-gas from the producer is first washed by water. The average composition of water-gas is by volume:—

	Liquid, B.P.	Per Cent.
Hydrogen - - - - -	−253° C.	48-54
Carbon monoxide - - -	−192° C.	42-44
Carbon dioxide - - - -	− 80° C.	5-2
Nitrogen - - - - -	−196° C.	5-3

The water-gas is then compressed, cooled, and treated with water under pressure, which removes practically all the carbonic acid, the remainder being absorbed by caustic soda. This dried compressed water-gas then passes through the temperature exchanger of the separator to the coil B (Fig. 24) immersed in a bath K containing liquid carbon monoxide.

In this coil the carbon monoxide is almost entirely condensed, and collects in the bottom of the liquefier A, whence it can be discharged by means of the cock R into K. The remaining gas passes through the liquefier A, surrounded by liquid nitrogen evaporating under vacuum. In this way more carbon monoxide is condensed, and the gas issuing at the top is composed of about 97 per cent. hydrogen. The remaining carbon monoxide can be removed by passage over soda lime.

After purification the gas contains:—

CO_2 - - - - - -	0 per cent.
Heavy hydrocarbons - - -	0 ,,
Oxygen - - - - - -	0 ,,
CO - - - - - -	0 ,,
Hydrogen - - - - -	99.2-99.4 ,,
Marsh gas - - - - -	0 ,,
Nitrogen - - - - -	0.8-0.6 ,,

with a specific gravity of .077 to .079.

The waste carbon monoxide gas can be used to generate power. In plants from 100 cub. m. per hour upwards the waste carbon monoxide is sufficient to drive the whole plant.

Size of Plant.	I.	II.	III.	IV.	V.
Cubic metres of hydrogen per hour - -	25	50	100	200	500
Water-gas required per hour - - -	70	125	250	500	1,250
Cubic metres cooling water required per hour -	2.25	3.80	7.60	13.50	32.50

Inventors in large numbers have applied themselves to the problem of the production of gases from mixtures by means of liquefaction, but the processes already described are still the only ones established on a large commercial basis. For details of other methods a perusal of the various patents (above) is recommended. Among others, which may contain the basis of a commercially sound process, may be mentioned those of Hampson, Thrupp, Pictet, Hildebrandt, and Levy. A patent by Levy and Hellbronner in 1902 applies the process of lixiviation by liquid rich in nitrogen by somewhat different means to the foregoing. The process consists of carrying out the fractional distillation under pressure in order to produce reliquefaction of the vapours, and thus obtaining rectification by washing with increasingly cold liquid.

Hildebrandt has produced a large number of patents, chiefly dealing with improvements of special parts of liquefying apparatus (see pp. 3 and 4).

PROPERTIES OF INDUSTRIAL GASES

The following table gives the properties of the chief liquefied gases:—

TABLE OF PHYSICAL CONSTANTS OF LIQUEFIED GASES (Dickerson).

Substance.	Symbol.	Critical Temperatures. Degs. Cent.	Critical Temperatures. Degs. Fahr.	Critical Pressure. Atmospheres.	Boiling Points under Atmospheric Pressure. Degs. Cent.	Boiling Points under Atmospheric Pressure. Degs. Fahr.	Freezing Point. Degs. Cent.	Freezing Point. Degs. Fahr.	Pressure at which Freezing Point was Determined. mm.	Density of Gas.	Density of Liquid at Temperature Given.	Colour of Liquid.
Water	H_2O	365	689	200	100	212	0	32	760	...	1 at 4° C.	Colourless
Ammonia	NH_3	130	266	115	−33	−27	−77	−107	...	8.5	0.6364 at 0° C.	,,
Propane	C_3H_8	97	206.6	44	−45	−49	Still liquid at −151° C.		...	20.95	...	,,
Acetylene	C_2H_2	37	98.6	...	−85	−121	−81	−113.8	950	12.97	...	,,
Nitrous oxide	N_2O	35	96	75	−89	−128	−115	−175	760	21.99	...	,,
Ethane	C_2H_6	34	93.2	50.2	−93	−135.4	Still liquid at −151° C.		...	19.97	...	,,
Carbon dioxide	CO_2	31	88	75	−80	−112	−56	−69	760	21.94	0.83 at 0° C.	,,
Ozone	O_3	...	...	...	−93	−135.4	...	...	...	23.89	...	Dark blue, easily exploded
Ethylene	C_2H_4	10	50	51.7	−102	−150	−169	−272	...	13.97	...	Colourless
Methane	CH_4	−81.8	−115.2	54.9	−164	−263.4	−185.8	−302.4	80	7.98	0.415 at −164° C.	,,
Nitric oxide	NO	−93.5	−135	71.2	−153.6	−254	−167	−369	138	14.98	...	,,
Oxygen	O_2	−118.8	−182	50.8	−181.4	−294.5	...	...	...	15.96	1.124 at −181.4° C.	Blue
Argon	A	−121	−185.8	50.6	−187	−304.6	−189.6	−309.3	...	19.9	about 1.5 at −187° C.	Colourless
Carbon monoxide	CO	−139.5	−219.1	35.5	−190	−310	−207	−340.6	100	13.96	...	,,
Air	...	−140	−220	39	−191.4	−312.6	...	...	...	...	0.933 at −191.4° C.	Light blue
Nitrogen	N_2	−146	−231	35	−194.4	−318	−214	−353.2	60	14.01	0.885 at −194.4° C.	Colourless
Hydrogen	H_2	−234	−389	20	−243	−405	...	..	...	1	...	,,
Helium	He	...	...	...	Below −264	−443.2	...	...	...	2.02	...	,,

CHAPTER II

Industrial Oxygen

CHAPTER II

INDUSTRIAL OXYGEN

BY GEOFFREY MARTIN, D.Sc., Ph.D.

Manufacture.—The main method is **from liquid air** in the manner described in the preceding article by Mr Dickson.

The following processes may be mentioned in addition to the method already described:—

1. **The Brinn Process.**—By means of barium peroxide. No longer worked.
2. **The Jaubert Process.**—Potassium permanganate is mixed with sodium peroxide. When water is added oxygen gas is spontaneously evolved. Somewhat cheaper is the process (patented in 1902) of mixing together and compressing into tablets sodium peroxide and bleaching powder (100 parts of bleaching powder with available Cl = 33-35 per cent., to 39 parts of Na_2O_2). On bringing this into contact with water, oxygen is evolved according to the equation:—

$$CaOCl_2 + Na_2O_2 = CaO + 2NaCl + 2O.$$

Convenient for use on a small scale. Sold under name "Oxygen Cubes."

3. **Electrolytic Oxygen.**—This is obtained as a by-product in the manufacture by electrolysis of caustic soda, etc. (see **Martin's** "Industrial Chemistry," Vol. II.), hydrogen being also obtained at the same time.

It can, however, only be produced economically at places where electrical power is very cheap.

Properties of Oxygen.—The main physical properties of oxygen are set forth in the table on page 23.

Its main characteristic is the intensity with which it supports combustion.

Uses of Industrial Oxygen.—The main use of industrial oxygen is the autogenous welding and cutting of metals.

To give some examples—the ease and rapidity with which the most extensive and massive iron structures can be cut through in almost any position by a few workmen in a few minutes, work which in the ordinary way can only be effected by many workmen in a great many days. Its use for demolishing iron structures, etc., has been demonstrated in innumerable cases. The cutting out of a manhole in a boiler of average size will employ two workmen seven to eight hours. By means of oxygen the work can be done in about eight minutes.

Industrial oxygen is at present principally employed:—

1. **In the Motor Car Industry.**—For welding together autogenously metals like cast iron, steel, magnalium, aluminium, and for effecting rapid repairs of broken parts.
2. **For Illuminating Purposes,** a jet of oxygen being impinged on lime, etc.
3. **In Mining and Tunnelling and Iron Smelting,** *e.g.*, stoppages in blast ovens are often melted through.
4. **In the Tinning and Enamelling Industry,** for melting and welding different metals.
5. **In the Manufacture of Varnishes, etc.,** manufacture of SO_3, etc.
6. **The Gas Industry.**—Enriching the air with oxygen allows the smokeless burning of certain fuels in furnaces, also the burning of wet peat, wood waste, etc., in the production of combustible or producer-gas, water-gas, etc.
7. **The Glass and Ceramic Industry** use oxygen for cutting through glass plates, cylinders, etc., also for increasing the temperature of certain small furnaces.
8. **Medicinal use of Oxygen.**—For saving life for "gassed" workmen, and for administrating to those suffering from lung complaints.
9. **In Lead Work,** for making leaden vessels, sulphuric acid chambers, etc.

Innumerable other uses are growing up for oxygen gas, but the above suffices to give the reader an idea of the enormous extent to which oxygen is now being employed in industry.

Statistics.—Germany in 1905 used only about 40,000 cub. m. of oxygen.
The following table shows the amount of oxygen supplied from Linde plants *alone* in 1911 :—

	Cub. Metres.		Cub. Metres.
Germany	6,000,000	Sweden	80,000
England	2,230,000	Roumania	80,000
France	1,550,000	Russia	80,000
Italy	680,000	America	2,676,000
Austria-Hungary	660,000	India	144,000
Spain	220,000	China	80,000
Switzerland	180,000	Australia	220,000
Denmark	110,000		

CHAPTER III

Industrial Nitrogen

CHAPTER III

INDUSTRIAL NITROGEN

BY GEOFFREY MARTIN, Ph.D., D.Sc.

Manufacture.—Nitrogen is now manufactured almost exclusively from atmospheric air by the liquefaction process, as described in Chapter I. on the **Liquefaction of Gases.**

In some Continental works the nitrogen is produced by passing air over hot

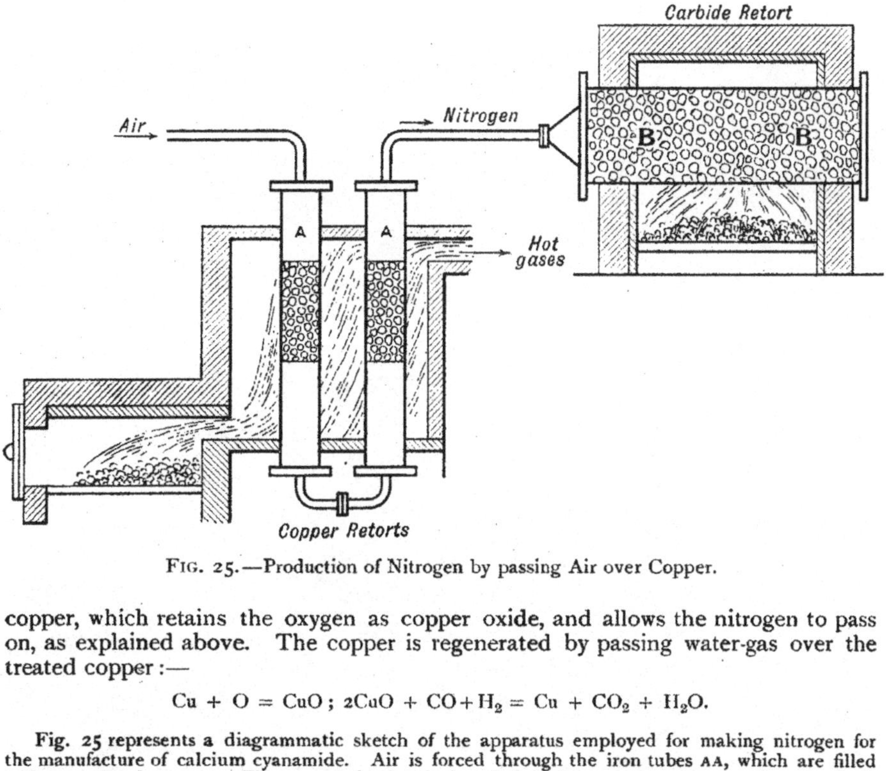

FIG. 25.—Production of Nitrogen by passing Air over Copper.

copper, which retains the oxygen as copper oxide, and allows the nitrogen to pass on, as explained above. The copper is regenerated by passing water-gas over the treated copper:—

$$Cu + O = CuO ; 2CuO + CO + H_2 = Cu + CO_2 + H_2O.$$

Fig. 25 represents a diagrammatic sketch of the apparatus employed for making nitrogen for the manufacture of calcium cyanamide. Air is forced through the iron tubes AA, which are filled with granulated copper. The oxygen is absorbed and the nitrogen passes on, to be absorbed, say, in the calcium carbide in the retort B.

Crude nitrogen, contained in the waste gases from furnaces, is also used for making aluminium nitride (see **Martin's** "Industrial Chemistry," Vol. II.).

Properties.—Pure nitrogen is a colourless, tasteless, and inodorous gas. It does not support combustion, it is non-inflammable, and does not turn lime-water milky (distinction from carbon dioxide). It combines directly with substances such as calcium, barium, magnesium, titanium, silicon, etc., forming compounds known as "nitrides." Chemically pure nitrogen possesses the following properties:—Critical temperature, −146° C.; critical pressure, 35 atmospheres; boiling point at atmospheric pressure, −194.4° C.; freezing point, −214° C.; density of gas, 14.01 (H = 1), or 0.96717 (air = 1); specific gravity of liquid nitrogen at its B.P. is 0.8042; solid nitrogen has a specific gravity of .0265. The weight of a litre is 1.25107 g. (Rayleigh). Atmospheric nitrogen contains argon, and so its density is somewhat greater than that of chemically pure nitrogen, the density of "atmospheric" nitrogen being 0.97209, and one litre at 760 mm. weighs 1.25718 g. (Rayleigh).

Uses.—Nitrogen is now being manufactured on an enormous scale for the production of **Synthetic Ammonia** (which see below), calcium cyanamide (nitrolim) (see **Martin's** "Industrial Chemistry," Vol. II.). It is also used for making aluminium nitride, and other nitrides (see Serpek process, **Martin,** *loc. cit.*).

When nitrogen gas (or ammonia or even nitrogen-containing substances) is brought into contact with white-hot iron, some nitrogen is absorbed by the surface, and an extremely hard surface is thus produced (case-hardening).

The nitrogen gas industry is as yet in its infancy, but is rapidly developing.

CHAPTER IV

Hydrogen

CHAPTER IV

HYDROGEN

BY H. STANLEY REDGROVE, B.Sc.(Lond.), F.C.S.

PROPERTIES

HYDROGEN is a colourless, odourless, and tasteless gas. It is the lightest element known, its atomic weight (O = 16) being 1.0008, and its density, compared with air as unity, being 0.06949. It has been liquefied by **Dewar** and others, and boils at −252.5° C. Liquid hydrogen is a clear, colourless fluid, part of which may be solidified by rapid evaporation of the rest. Solid hydrogen is also colourless, and has a melting point of −257° C., according to **Dewar**, or −258.9° C., according to **Travers**.

Hydrogen is a very widespread element, since it forms about one-ninth part by weight of water. It occurs free in nature in the gas of certain volcanoes, and in combination with other elements in petroleum and most organic bodies, as well as in water.

Hydrogen combines vigorously with oxygen to form water, according to the equation:—

$$2H_2 + O_2 = 2H_2O.$$

Mixtures of hydrogen and oxygen combine with explosion when ignited. Hydrogen burns readily in the air, with an almost colourless flame (faintly green, tinged with violet-blue). The colour may be altered by changing the pressure. The heat evolved in the combustion of hydrogen is very great. At constant pressure and temperature $t°$ C., the heat liberated per gram-molecule of water produced is $57.9 + 0.00165t$ Cals., when the water is in the gaseous state, or $68.5 - 0.0077t$ Cals., when, the water being liquefied, its latent heat is also obtained.[1]

Hydrogen will not support the combustion of bodies which we usually regard as combustible, nor will it support life. It is not, however, poisonous, and certain bodies, for instance oxygen and chlorine, will burn in it.

It is very slightly soluble in alcohol, and still less soluble in water.

Certain metals, palladium in particular, have the remarkable property of absorbing large quantities of hydrogen at high temperatures and pressures, with very little alteration in their physical properties and appearance. Such "occluded" hydrogen, as it is called, exhibits increased chemical activity. The hydrogen is evolved when the metal containing it is placed under reduced pressure. Palladium will absorb over 900 times its volume of hydrogen when heated in an atmosphere of the gas, whilst black platinum will absorb about 100 volumes under similar conditions.

Recent experiments have shown that certain bodies (for example, unsaturated fats) will readily enter into combination with hydrogen if palladium, black platinum, or especially nickel, or an oxide of nickel, is present. This phenomenon is, no doubt, closely connected with that of occlusion, but its exact nature, whether the hydrogen chemically combines with the metal to form a definite compound, or merely "alloys" with it in the manner that two metals may be alloyed, is still a matter of debate. In its combinations hydrogen certainly appears to be capable of acting both as a metal and as a non-metal.

MANUFACTURE

Water is by far the cheapest raw material for the manufacture of hydrogen. Next in point of cheapness come hydrocarbons. There are a few processes making use of other sources of hydrogen, but they are in all cases much more expensive, and are either being superseded, or have only a limited application for special purposes. The various processes may be roughly classified as follows:—

1. By the Action of Water on Metals.
2. By the Action of Water on Non-Metals.
3. By the Deoxidising Action of Carbon Monoxide.
4. By the Electrolysis of Dilute Sulphuric Acid, Alkali, or certain Salt Solutions.
5. From Hydrocarbons.
6. By the Action of Acids on Metals.
7. By the Action of Alkalis on Metals.
8. Special Methods.

These will now be considered in detail.

1. By the Action of Water on Metals.

The alkali metals (*e.g.*, sodium) and the metals of the alkaline earths (*e.g.*, calcium) decompose water at ordinary temperatures, the first violently and the second readily, as under:—

$$2Na + 2H_2O = 2NaOH + H_2.$$
$$Ca + 2H_2O = Ca(OH)_2 + H_2.$$

Certain metals which, in a pure condition, will attack water only at high temperatures, will react at lower temperatures if alloyed with some other metal, *e.g.*, zinc, if first coated with a thin layer of copper, will readily decompose warm water ($Zn + 2H_2O = Zn(OH)_2 + H_2$).

These processes are only applicable for the preparation of pure hydrogen in the laboratory. A somewhat similar process has, however, been patented by **Uyeno** (English Patent, 11,838, 1912; United States Patent, 1,118,595, 1914), in which an alloy of aluminium with small proportions of zinc and tin, together with a trace of mercury, is used to decompose hot water. The method seems well adapted to the preparation of hydrogen for aeronautical purposes; but such an alloy would be too expensive for manufacturing hydrogen for industrial use, and iron is practically the only metal used industrially for the decomposition of water.

When steam is passed over hot iron, in a suitable furnace, hydrogen is produced in accordance with the equation—

$$3Fe + 4H_2O = Fe_3O_4 + 4H_2.$$

The iron can be revivified by reducing the oxide formed by means of gas containing carbon monoxide (for example, producer-gas, *q.v.*), thus:

$$Fe_3O_4 + 4CO = 3Fe + 4CO_2.$$

This is **Lane's** process.

The two reactions are carried out alternatively so long as they are found to take place with sufficient speed, ten minutes being allowed for the first ("gassing" or oxidation period), and about twice as long for the second (reduction period). A temperature of from about 900°-1,000° C. is used for both. In order to make the process practically continuous, two or more furnaces may be run alternatively (see **Lane's** Patents, English Patents, 10,356, 1903; 17,591, 1909; 11,878, 1910, etc.).

The process has been found very efficient when a suitable type of iron is employed, though there are a good many technical difficulties connected with it, which a number of patents taken out during the last few years claim to have obviated. Spongy iron appears to be far more reactive than the compact form, which is apt to become only superficially oxidised (**Badische Anilin und Soda Fabrik**, English Patent, 27,735, 1912; French Patent, 453,077, 1913; **Anton Messerschmitt**, United States Patent, 971,206, 1908). However, compact iron will decompose steam, and **H. Dicke** (German Patent, 280,964, 1913) proposes that the scrap-iron that accumulates in iron and steel works be utilised for this purpose. **Messerschmitt** (English Patent, 18,028, 1913; French Patent, 461,480, 1913; United States Patent, 1,109,448, 1914) states that good yields of hydrogen can be obtained at about 700°-800° C. by using a mixture of iron and manganese produced by the reduction of ores containing both these metals; whilst **W. Näher** and **M. Nöding** (German Patent, 279,726, 1913) pass the steam first over copper and then over iron.

Messerschmitt is responsible for a large number of improvements in apparatus, etc., for the manufacture of hydrogen by the deoxidation of steam by means of iron, for which see German Patents, 263,391, 268,062, 268,339, 1912; French Patents, 444,105, 444,106, 1912; 461,623, 461,624, 461,625, 462,379, 1913; English Patents, 12,117, 12,242, 12,243, 1912; 17,690, 17,691, 17,692, 18,942, 1913; United States Patent, 1,109,447, 1914. For improvements in apparatus the following patents should also be consulted—**Badische Anilin und Soda Fabrik**, English Patent, 2,096, 1913; and **Berlin Anhaltische Maschinenbau A.-G.**, English Patents, 28,390, 1913; 6,155, 1914; French Patent, 465,474, 1913; United States Patents, 1,118,595, 1914; 1,129,559, 1915. Purification of the reducing gas employed is essential in order that the efficiency of the iron shall be preserved. With due precautions the hydrogen obtained is in an almost chemically pure condition, containing only traces of nitrogen. According to **Bergius** ("Production of Hydrogen from Water and Coal from Cellulose at High Temperatures and Pressures," *The Journal of the Society of Chemical Industry*, Vol. XXXII., 1913, p. 462; German Patent, 254,593, 1911; French Patent, 447,080, 1912; English Patents, 19,002, 19,003, 1912; United States Patents, 1,059,817, 1,059,818, 1913), water is more reactive in the liquid state than in the gaseous state. Consequently he carries out the reaction between water and iron at a sufficiently high pressure to prevent the vaporisation of the water. The resulting hydrogen is particularly pure, owing to the fact that under the conditions of the method the water does not attack the impurities in the iron. One specimen of hydrogen obtained by this method gave the following results on analysis:—

Hydrogen	99.95 per cent.
Carbon monoxide	0.001 ,,
Saturated hydrocarbons	0.042 ,,
Unsaturated hydrocarbons	0.008 ,,

The speed of the reaction rapidly increases with a rise in the temperature, and it is facilitated by the presence of a trace of an electrolyte, such as ferrous chloride, and of a metal more electro-positive than iron, as shown by the following table:—

	Temperature, 0° C.	Volume of Hydrogen Evolved per Hour.
		c.cs.
Iron and pure water -	300	230
,, $+ FeCl_2$ - -	300	1,390
,, $+ FeCl_2 + Cu$ -	300	1,930
,, $+ FeCl_2 + Cu$ -	340	3,450

The iron oxide (Fe_3O_4) is obtained in the form of a finely divided powder, which can be readily reduced by means of carbon or carbon monoxide. Another advantage of the method lies in the fact that the hydrogen is obtained under sufficiently high pressure for filling into cylinders without further compression. The process is worked at a factory in Hanover, and it is claimed that the method is so rapid that a generating vessel of 10 gallons will produce about 3,000 cub. ft. of hydrogen per day at a cost of about ¾d. per cubic metre.

2. By the Action of Water on Non-Metals.—Certain of the non-metals can be made to react with water so as to yield an oxide and free hydrogen. The only one of industrial importance is carbon. When steam acts on heated coke the following reactions take place:—

$$C + H_2O = CO + H_2,$$
$$C + 2H_2O = CO_2 + 2H_2.$$

The mixture of carbon monoxide and hydrogen obtained in the first case is known as "Water-Gas" (see the present writer's article on "Producer-Gas" in this volume). Efforts have been made to cause the second of these reactions to take place exclusively. **O. Dieffenbach** and **W. Moldenhauer** (English Patent, 8,734, 1910) claim that this may be accomplished by impregnating the coke with 10 per cent. of potassium carbonate, and mixing it with five times its weight of lime, which absorbs the carbon dioxide produced. A temperature of 550°-750° C. is employed. **F. Bergius** (German Patent, 259,030, 1911) uses a sufficiently high pressure to prevent the vaporisation of the water (see above). In these circumstances he states that the formation of carbon monoxide may be prevented by the use of certain catalytic agents, of which he prefers thallium chloride (1 per cent. of the weight of the coke). He uses a temperature of 340° C., and the carbon dioxide produced is absorbed by lime.

The manufacture of hydrogen from water-gas is now one of considerable importance, and several methods have been proposed for the separation of the carbon monoxide from the hydrogen.

Thus **A. Frank** (French Patent, 371,814, 1906; English Patent, 26,808, 1906) suggested the use of cuprous chloride solution, which absorbs most of the carbon monoxide. The dried gas is then passed over calcium carbide at a temperature exceeding 300° C., when the following reactions take place:—

$$CaC_2 + CO = CaO + 3\,C,$$
$$2CaC_2 + CO = 2CaO + 5C.$$

The carbon dioxide and nitrogen present are also absorbed, with the production of calcium carbonate and calcium cyanamide respectively. The process, however, proved too expensive to work for industrial purposes, which led to further investigation, resulting in the Linde-Frank-Caro process, dealt with below.

Physical methods have also been proposed for the separation of water-gas into its constituents. Centrifugation (see English Patent, 12,194, 1902, **E. N. Mazza**) has not proved satisfactory (see also English Patent, 17,946, 1905; **C. Clamond**). Diffusion has also been proposed, **Jouvier** and **Gautier** (French Patent, 372,045, 1906) claiming that a single passage through a partition of unglazed porcelain reduces the amount of carbon monoxide by nearly 82 per cent.

In consequence of the great difference between the boiling point of hydrogen (−252.5° C.) and the boiling points of carbon monoxide (−190° C.) and nitrogen (−195.5° C.), liquefaction has proved to be by far the most important method industrially for the separation of hydrogen from the carbon monoxide of water-gas, and the nitrogen usually present as an impurity. **Linde's** work, cited above, should be specially consulted. The main principle in **Linde's** method for the liquefaction of gases is that of expansion without external work at low temperatures and high pressures, in which circumstances the Thomson-Joule effect becomes sufficient to produce a large lowering in the temperature. This method is combined with the use of counter-current interchangers. In the **Linde-Frank-Caro** process[1] for the separation of water-gas, worked by the **Ges. für Lindes Eismaschinen,** water-gas, having approximately the composition H_2 48-54 per cent., CO 42-44 per cent., CO_2 2-5 per cent., N_2 3-5 per cent., with traces of other impurities, is first freed from its carbon dioxide by means of water under pressure, a final purification being effected by means of alkalis. It is then dried and passed into a special cooling apparatus. Most of the carbon monoxide is liquefied by passing the gas through a coil in a vessel containing liquid carbon monoxide already produced by the process. The liquid carbon monoxide thus formed is expanded and poured into this vessel. The hydrogen passes on, and a further separation of carbon monoxide and nitrogen is effected by cooling the gas in boiling nitrogen. The separated hydrogen and carbon monoxide gases issuing from the apparatus pass through a counter-current interchanger, and thus facilitate the cooling of the water-gas which enters the apparatus through the interchanger.

The hydrogen thus obtained is 97-97.5 per cent. pure, the impurities being carbon monoxide 1.7-2 per cent., nitrogen 1-1.8 per cent. Other impurities are practically nil. The cost of production is 1½d. per cubic metre. The gas may be further purified, at an additional cost of ½d. per cubic metre, by means of soda-lime, which removes the whole of the remaining carbon monoxide, giving a product containing 99.2-99.4 per cent. hydrogen, the remainder being nitrogen. The gas is claimed to be free from such impurities as would interfere with its use in aeronautics or in the synthesis of ammonia. Further, it is obtained under pressure, which is important for the latter process. The carbon monoxide obtained may be used as a fuel.

In **Claude's** process for the separation of the constituents of water-gas by means of refrigeration (French Patents, 329,839, 1903; 469,854, 1913; English Patent, 13,160, 1914) the main point is that cooling is effected by expansion of the hydrogen with external work in a motor lubricated with light petroleum.

See also **Jaubert's** auto-combustion process under "Special Methods" below.

3. By the Deoxidising Action of Carbon Monoxide.—Other processes for the manufacture of hydrogen from water-gas utilise the deoxidising action of carbon monoxide, in virtue of which the latter gas can be absorbed by suitable bodies and replaced by an equal volume of hydrogen. Thus the **Chemische Fabrik Greisheim-Elektron** have patented a process (English Patents, 2,523, 1909, and 13,049, 1912) in which hydrogen is produced by means of the following reaction:—

$$Ca(OH)_2 + CO = CaCO_3 + H_2.$$

The water-gas, together with steam, is passed over lime at a temperature of about 500° C., which is below that of the dissociation of calcium carbonate. As the process is exothermic, extraneous heating is not required once it is started. 5 per cent. of powdered iron is added as a catalyst. Calcium oxide is regenerated from the carbonate produced. The cost of the hydrogen, containing as an impurity the nitrogen present in the water-gas, is stated to be about 1d. per cubic metre. It is probable that the reaction takes place in two stages, calcium formate being first produced (see below). **La Soc. l'Air Liquide** makes use of the same reaction in connection with **Claude's** liquefaction process (French Patent, 453,187, 1912;

[1] See Chapter I. by Mr Dickson on "The Liquefaction of Gases," p. 21.

English Patent, 7,147, 1913; United States Patent, 1,135,355, 1915). (See also **H. Strache,** German Patent, 253,705, 1910.)

A number of patents based on the reaction $CO + H_2O = CO_2 + H_2$ have been taken out in recent years:—

The **Badische Anilin und Soda Fabrik** (English Patents, 26,770, 27,117, 27,955, 1912; 8,864, 1913; 16,494, 1914; French Patent, 459,918, 1913) employ as catalysts finely divided iron (prepared below 650° C.), nickel, or similar bodies. The gases are kept under a pressure of 4 to 40 atmospheres, and the temperature, which is a factor of importance, is maintained at about 400°-500° C. The requisite heat may be obtained by the admission of oxygen. The carbon dioxide is removed by absorption. In German Patent, 279,582, 1913, they state that the addition of oxygen compounds of chromium, thorium, uranium, beryllium, antimony, etc., to the iron increases its reactivity as a catalyst (see also German Patent, 282,849, 1913). **Buchanan** and **Maxted** (English Patent, 6,477, 1914) suggest the use as a catalyst of iron-copper couple made by the reduction of copper nitrate by means of reduced iron (see also English Patent, 6,476, 1914).

In this connection the formation of hydrogen as a by-product in the manufacture of oxalates may be mentioned. Oxalates are now largely manufactured from formates, the latter being obtained by the action of carbon monoxide (producer-gas) on alkalis under pressure, thus:—

$$KOH + CO = \underset{\text{Pot. formate.}}{H.COOK}$$

$$2H.COOK = \underset{\text{Pot. oxalate.}}{\begin{matrix} COOK \\ | \\ COOK \end{matrix}} + H_2$$

(See United States Patent, 1,098,139, 1914; *cf.* **Badische Anilin und Soda Fabrik**, English Patent, 1,759, 1912.) A full description of manufacture is given in **Martin's** "Industrial Chemistry," Vol. I., pp. 369-370.

4. **By Electrolysis.**—When an electric current is passed through water containing a small quantity of sulphuric acid or an alkali, the water is decomposed into its elements, the oxygen being liberated at the anode or positive pole, and the hydrogen escaping at the kathode or negative pole. The process is sometimes called "the electrolysis of water," but the expression is a misnomer, since the acid or alkali plays an essential part in the reaction, pure water being practically a non-electrolyte and a non-conductor of electricity. The changes involved may be shown thus:—

With sulphuric acid - - -	$H_2SO_4 = \overset{+}{H} + \overset{+}{H} + \overset{--}{SO_4}$.
At the kathode - - -	$\overset{+}{H} + \overset{+}{H} = H_2$.
At the anode - - -	$2\overset{--}{SO_4} + 2H_2O = 2H_2SO_4 + O_2$.
With caustic potash - - -	$KOH = \overset{+}{K} + \overset{-}{OH}$.
At the kathode - - -	$2\overset{+}{K} + 2H_2O = 2KOH + H_2$.
At the anode - - -	$2\overset{-}{OH} + 2\overset{-}{OH} = 2H_2O + O_2$.
With barium hydroxide - -	$Ba(OH)_2 = \overset{++}{Ba} + \overset{-}{OH} + \overset{-}{OH}$.
At the kathode - - -	$\overset{++}{Ba} + 2H_2O = Ba(OH)_2 + H_2$.
At the anode - - -	$2\overset{-}{OH} + 2\overset{-}{OH} = 2H_2O + O_2$.

By using a platinum plate as kathode, and a platinum wire immersed in an amalgam of zinc and mercury (to absorb the oxygen) as anode, very pure hydrogen may be obtained in the laboratory, especially if thoroughly purified barium hydroxide is used (**Baker**, *Journal of the Chemical Society*, Vol. LXXXI., 1902, p. 400).

Under manufacturing conditions, hydrogen produced by the electrolysis of dilute sulphuric acid and alkalis usually contains oxygen (1-2 per cent.), an impurity liable to produce dangerous explosions in certain of the industrial applications of hydrogen (*e.g.*, the synthesis of ammonia). To obviate this, diaphragms are introduced between the electrodes, though this increases the electrical resistance of the cells. If metallic diaphragms are employed they are pierced to allow the passage of the ions. Alternatively, asbestos diaphragms may be used. In the **Garuti** and **Pompili** apparatus, used largely in Italy (English Patents, 23,663, 1896; 12,950, 1900; 2,820, 1902; 27,249, 1903), the diaphragms consist of thin iron sheets, the perforations in which are covered both sides with a metallic net. More recent types of apparatus will be found described in French Patents, 397,319, 1908; 459,967, 1912; 461,981, 472,394, 463,507, 1913; United States Patents, 981,102, 1,003,456, 1,004,249, 1911; 1,035,060, 1912; 1,086,804, 1,087,937, 1,094,728, 1914, etc.

On electrolysing potassium chloride solution, caustic potash, chlorine, and hydrogen are obtained. Thus:—

$$KCl = \overset{+}{K} + \overset{-}{Cl}.$$

At the kathode - - - $2\overset{+}{K} + 2H_2O = 2KOH + H_2$.

At the anode - - - $\overset{-}{Cl} + \overset{-}{Cl} = Cl_2$.

Similar reactions take place in the case of sodium chloride. Consequently enormous amounts of hydrogen are liberated in the manufacture of caustic alkalis by the electrolytic method (for particulars of which see **Martin's** "Industrial and Manufacturing Chemistry," Vol. II.), and, in Germany especially, the manufacture of alkali by that process is advantageously combined with that of chlorine and hydrogen, all of which are important articles of commerce. **Crossley** (*loc. cit.*) estimates that, in Germany, the electrical energy required for the process costs about ⅜d. per cubic metre of hydrogen.

5. **By the Decomposition of Hydrocarbons.**—Under certain conditions acetylene undergoes decomposition into its elements, thus:—

$$C_2H_2 = 2C + H_2 + 47.8 \text{ Cals.},$$

the reaction yielding not only hydrogen, but also carbon in a fine state of division, which may be used as pigment. The **Carbonium Gesellschaft** of Fredericks-haven produce the decomposition electrically, and the hydrogen thus obtained is used for filling Zeppelin dirigibles. Other hydrocarbons can be made to decompose into hydrogen and carbon, but as the reaction is usually endothermic so far as concerns hydrocarbons available for industrial purposes, a constant supply of energy is necessary. The reaction may be brought about by means of a very high temperature, maintained by means of external heating, but methods based upon this fact suffer from the disadvantage of heavy wear and tear of the plant, occasioned by the high temperature needed.

R. Lessing (English Patent, 15,071, 1909) suggests that hydrocarbon gases be decomposed in retorts at 1,000°-1,300° C., whilst R. P. Pictet (English Patent, 13,397, 1911) decomposes the vapours obtained by the distillation of petroleum oils by heating them to 1,350° C. In an earlier patent (English Patent, 24,256, 1910; see also French Patent, 421,838, 1910; German Patent, 255,733, 1912) the latter utilises the heat obtained by the decomposition of acetylene. Since this decomposition is exothermic, once initiated it continues automatically, and the heat thus obtained is used to decompose hydrocarbons (*e.g.*, benzine) whose decomposition is endothermic. The reaction is carried out by passing a mixture of acetylene with the other hydrocarbon through a conduit, the temperature of which is maintained at 500° C., at a pressure slightly above the normal.

The vapours of hydrocarbons (*e.g.*, petroleum, benzine, etc.) may also be decomposed by the action of steam at a high temperature, when hydrogen and carbon monoxide are obtained (**Pictet**, English Patent, 14,703, 1911; United States Patent, 1,134,416, 1913). The **Badische Anilin und Soda Fabrik** (French Patent, 463,114, 1913; English Patent, 12,978, 1913; United States Patent, 1,128,804, 1915) pass a mixture of hydrocarbons and steam over an inactive, refractory oxide, such as magnesia, covered with nickel oxide, or some other suitable catalyst, at a temperature of 800°-1,000° C. The resulting hydrogen is suitable for the production of ammonia after being freed from oxides of carbon.

Coke may also be used for the decomposition of hydrocarbons. Thus the **Berlin Anhaltische**

Maschinenbau A.-G. (German Patent, 267,944, 1913; French Patent, 466,040, 1913; English Patent, 2,054, 1914) obtain gases suitable for heating or lighting, and, under suitable conditions, hydrogen, by passing vaporised petroleum oils over incandescent coke, whilst **C. Ellis** (United States Patent, 1,092,903, 1914) treats coke, rendered white hot by an air blast, with heavy petroleum oil. The gas produced during the blowing period, and the hydrogen formed during the gassing stage, are removed by means of steam passed over, but not through, the bed of coke.

6. **By the Action of Acids on Metals.**—Certain acids, *e.g.*, dilute sulphuric and hydrochloric acids, readily generate hydrogen by their action on certain metals, for instance iron, zinc, and magnesium. Thus:—

$$Fe + H_2SO_4 = FeSO_4 + H_2.$$
$$Zn + 2HCl = ZnCl_2 + H_2.$$
$$Mg + H_2SO_4 = MgSO_4 + H_2.$$

These methods are employed in the laboratory to prepare hydrogen. The last is used when a very pure gas is required. The earliest industrial methods of manufacturing hydrogen made use of the action of sulphuric acid on such metals as scrap iron, zinc distillation residues, etc., and a very large number of patents have been taken out (*e.g.*, English Patents, 16,277, 1896; 15,509, 25,084, 1897; 17,516, 1898; 6,075, 18,930, 1905; 21,440, 1906); but such processes are now practically obsolete owing to the enormous quantity of material requisite. Thus, as **Crossley** points out (*loc. cit.*), a small military airship of 250,000 cub. ft. capacity would necessitate the use of 18 tons of iron and 31 tons of sulphuric acid to inflate it.

(*Cf.* also English Patent, 5,577, 1911, **G. Wannschaff** and **J. Savelsberg**, hydrogen and zinc oxide obtained by the action of zinc chloride solutions on waste zinc.)

7. **By the Action of Alkalis on Metals.**—Certain metals, *e.g.*, aluminium and zinc, have the power of liberating hydrogen from solutions of the caustic alkalis, thus:—

$$2Al + 6KOH = Al_2(OK)_6 + 3H_2.$$
$$Zn + 2KOH = Zn(OK)_2 + H_2.$$

Very pure hydrogen may be obtained in the laboratory by either of these processes. In the case of aluminium the evolution of hydrogen is very rapid. (*Cf.* **Uyeno's** patent above.)

Silicon behaves similarly with caustic alkalis, and **G. F. Jaubert** (French Patent, 45,4616, 1912; English Patent, 7,494, 1913) proposes to obtain hydrogen by the interaction of powdered silicon, aluminium, or their alloys, with an emulsion of a concentrated solution of caustic soda or potash with a non-saponifiable oil or grease, such as paraffin. See also English Patent, 17,589, 1911, and the same author's "Hydrogenite" process dealt with in "Special Methods" below.

8. **Special Methods.**—**Jaubert** has patented several ingenious processes by which hydrogen may be very rapidly and easily generated for military purposes. In French Patent, 327,878, 1902, he describes the preparation of calcium hydride. The hydrogen which is evolved as a by-product in many processes, such as the electrolytic production of soda (see above), is freed from oxygen, dried, and passed over pieces of calcium heated in an iron tube in an electric furnace. When brought into contact with water the calcium hydride thus obtained evolves hydrogen with violence, thus:—

$$Ca + H_2 = CaH_2.$$
$$CaH_2 + 2H_2O = Ca(OH)_2 + 2H_2.$$

The 90 per cent. pure calcium hydride prepared by the above method is known as "Hydrolith." One kg. of it yields 1 cub. m. of hydrogen, the evolution of the gas being so rapid that an average sized dirigible can be filled in about four hours. It is a very expensive method of generating hydrogen, but on account of its extreme rapidity has been adopted by the French army, and, as **Fourniols** (*loc. cit.*) points out, the saving in the cost of transport effected by using "hydrolith" in place of cylinders of hydrogen more than compensates for the expense of this material. See also English Patent, 25,215, 1907.

Jaubert has also patented mixtures which, when heated at one point, react violently in a manner like "Thermit" with the evolution of hydrogen.

"Hydrogenite," for example, consists of five parts of ferro-silicon, containing

90-95 per cent. of silicon, twelve parts of caustic soda, and four parts of slacked lime. It may be ignited by means of a hot iron ball, in a special apparatus, and 1 kg. yields about 320 litres of hydrogen (English Patent, 153, 1911).

Jaubert's auto-combustion methods are also of interest (French Patents, 427,191, 1910; 438,021, 1911; English Patents, 9,623, 1911; 5,005, 1912; United States Patent, 1,099,446, 1914). According to the first, a mixture consisting of a combustible body able to decompose water at a high temperature, and an oxidising agent able to support its combustion, is kindled in a closed vessel in the presence of water. The latter may be obtained by introducing a hydrated body which will give up its water at the temperature of the reaction. Suitable mixtures, for instance, may be made from iron filings, potassium perchlorate, and slacked lime, or from ferro-silicon, litharge, and soda-lime. In the second method, water (in the form of steam) is the only oxidant, and a highly heated metal, metalloid, or alloy, preferably ferro-silicon, is used as a combustible, the reaction being started in a closed vessel with a pinch of gunpowder.

Teissier and **Chaillau** (French Patent, 447,688, 1912) suggest the manufacture of hydrogen by passing steam under pressure over a mixture of barium sulphide and manganous oxide. The resulting mass of barium sulphate and manganous oxide is converted into a mixture of barium sulphide and manganese dioxide by the action of a red heat, from which the mixture of barium sulphide and manganous oxide is regenerated by the action of a white heat. Thus:—

$$BaS + 4MnO + 4H_2O = BaSO_4 + 4MnO + 4H_2.$$
$$BaSO_4 + 4MnO = BaS + 4MnO_2.$$
$$BaS + 4MnO_2 = BaS + 4MnO + 2O_2.$$

INDUSTRIAL USES OF HYDROGEN

Hydrogen has recently assumed great industrial importance, chiefly owing to the creation of two entirely new industries—the synthetic production of ammonia, and the hydrogenation of fats. Aeronautics are responsible for another demand, and hydrogen finds some other important and useful applications.

1. **Use as a Combustible.**—Water-gas consists of hydrogen mixed with carbon monoxide, and semi-water-gas also contains hydrogen. Both of these are very important fuels (see the present writer's article on "Producer-Gas" in this volume). When hydrogen, undiluted with other gases, is burned with oxygen, a temperature as high as 2,800° C. can be reached.

The oxy-hydrogen flame has one application in the production of the Drummond limelight. The hot flame, impinging on a cylinder of burnt lime, heats it to incandescence, and produces a dazzling white light. The effect is almost equal to that of the electric arc. The flame has other applications, namely, for fusing and welding platinum, in working fused quartz, and in the production of autogenous joints in lead and other metals. Such joints, that is to say, those formed by melting the metal itself for the purpose of welding, are found to be more durable than soldered joints.

The usual form of blow-pipe consists of a nozzle, fed by two tubes, fitted with stop-cocks. One tube leads from a hydrogen-generator or cylinder, the other runs to a bellows, or may be fed with oxygen, the best proportion of the gases being one volume of oxygen to four volumes of hydrogen. The oxy-acetylene flame, however, is now replacing the oxy-hydrogen flame, owing to the higher temperature produced by the former, and to the fact that less gas is required.

2. **Use as an Inert Atmosphere.**—Hydrogen, purified from traces of oxygen by passage over heated copper, may be used in certain processes which require an atmosphere free from oxygen, such as in the manufacture of tungsten filaments for electric lamps. These filaments have to be heated to a high temperature, when they are very sensitive to the action of oxygen. An atmosphere consisting of equal parts of hydrogen and nitrogen is usually employed.

3. **Use in Aeronautics.**—Owing to its extreme lightness, hydrogen is

superior to all other gases for balloons and dirigibles. The gas must be dried and freed from compounds of arsenic and phosphorus, as these latter are liable to produce arsenic and phosphoric acids, which may corrode the fabric of the balloon with disastrous results.

The lifting power of hydrogen prepared from zinc and sulphuric acid, and dried with lime, is given by **Walker** ("Aerial Navigation," 1902) at 68.5 lbs. per 1,000 cub. ft., whilst **Baden Powell** (paper read before The Royal United Service Institution, 1893) puts the figure at 60-70 lbs.

Specially rapid methods of preparing hydrogen for military purposes have been dealt with above.

4. **Use for the Synthesis of Ammonia.**—Ammonia is now largely manufactured synthetically from its elements ($N_2 + 3H_2 = 2NH_3$), which combine together with ease at high pressures and temperatures (500°-600° C.), in the presence of suitable catalysts, such as uranium containing uranium carbide. A full description of the processes is given in **Martin's** "Industrial Chemistry," Vol. II.

5. **Use for the Hydrogenation of Fats.**—Under suitable conditions hydrogen will combine with unsaturated organic compounds, thus:—

$$\underset{\text{(Oleic acid.)}}{C_{18}H_{34}O_2} + H_2 = \underset{\text{(Stearic acid.)}}{C_{18}H_{36}O_2}.$$

The saturated fatty acids and their glycerides have higher melting points than the corresponding unsaturated compounds, so that by this means various soft oils and fats, such as whale-oil, fish-oil, linseed-oil, etc., may be hardened, and their commercial value correspondingly increased. Moreover, the colour of such saturated fats and oils is rendered lighter by the process, and their characteristic odours and tastes are destroyed. The high melting-point products can be used for candle-making, soap-making, and in the manufacture of edible fats, etc. Consequently, enormous amounts of hydrogen are now used for the hardening of fats and oils. Full details of the processes are given in **Martin's** "Industrial Chemistry," Vol. I., p. 100*a*, *et seq.*

In order to carry out the reaction satisfactorily, attention must be paid to the temperature, and means must be taken to ensure intimate mixing of the oil and the hydrogen gas, preferably by spraying the oil into an atmosphere of hydrogen under pressure. The reaction takes place only in the presence of a suitable catalyst such as palladium, reduced nickel (prepared at a low temperature), or nickel oxide. The hydrogen must be free from impurities, sulphur being especially deleterious. The reaction is in many ways a mysterious one, and seems to be extremely sensitive to small details. (See English Patents, 1,515, 1903; 2,520, 13,042, 1907; 9,112, 1908; 3,752, 7,726, 18,969, 29,612, 30,014, 30,282, 1910; 8,147, 15,439, 15,440, 1911; 72, 4,702, 5,773, 18,282, 18,310, 18,998, 23,377, 23,643, 28,754, 1912; and "Reports on Patent Cases," 1913, Vol. XXX., p. 297; United States Patents, 1,004,034, 1,004,035, 1911; 1,026,156, 1,040,531, 1,040,532, 1,043,912, 1912; 1,058,738, 1,059,720, 1,060,673, 1913, etc.)

CHAPTER V

Producer-Gas

CHAPTER V
PRODUCER-GAS

By H. Stanley Redgrove, B.Sc.(Lond.), F.C.S.

CARBON MONOXIDE

The term "**producer-gas**" is applied to any gaseous fuel obtained from solid fuel in a "gas producer," in which the solid fuel is, as far as possible, completely consumed by means of partial oxidation, so as to give a combustible gas and leave no combustible residue. In actual practice the oxygen required for the combustion of the solid fuel in making producer-gas is obtained either from the air, from water, or (more frequently) from both sources. Hence three main varieties of producer-gas may be distinguished—**air-gas, water-gas,** and **semi-water-gas**—and as the essential component of each of these gases is carbon monoxide (or carbonic oxide), CO, some information may first be given concerning this gas.

Carbon Monoxide is obtained when carbon, or carbonaceous material, is burnt in a supply of oxygen or air insufficient to convert it into the dioxide, according to the equation—

$$2C + O_2 = 2CO.$$

The reaction is very frequently considered to take place in two stages. Firstly, according to this view, part of the carbon is converted into the dioxide, which then reacts with the remaining carbon to produce the monoxide, thus:—

$$C + O_2 = CO_2,$$
$$CO_2 + C = 2CO.$$

On the other hand, **Dixon**[1] has found that small quantities of the monoxide are produced by the action of oxygen on carbon at 500° C., whereas the reduction of carbon dioxide by carbon does not appear to take place below 600° C. Further, **Baker**[2] has found that perfectly dry oxygen will combine with carbon only with difficulty, and that the chief product is then the monoxide. Possibly both reactions take place simultaneously between carbon and oxygen, as indicated by the more recent work of **Rhead** and **Wheeler.** At any rate a state of equilibrium is established through the reversible reaction—

$$CO_2 + C \rightleftarrows 2CO.$$

This reaction has been fully studied by **Rhead** and **Wheeler**, who found, in accordance with **Le Chatelier's** theorem, that—

$$\frac{19,500}{T} + \log_e P + \log_e \frac{C_1^2}{C_2} = \text{a constant,}$$ [3]

where T is the temperature, P the pressure, C_1 the concentration of CO, and C_2 the concentration of CO_2. This formula shows that both increase in temperature and decrease in pressure favour the formation of the monoxide, which is only to be expected from the fact that the reduction of carbon dioxide by carbon is accompanied by the absorption of 39 Cals. per gram-molecule of dioxide.

Carbon monoxide is a colourless, tasteless, and almost inodorous gas.[4] It is almost insoluble in water, but is readily absorbed by an acid or ammoniacal solution of cuprous chloride.

Carbon monoxide readily burns in oxygen or air with a lambent blue flame, forming the dioxide. A trace of moisture, however, must be present, such as is always the case with ordinary air, otherwise combustion will not take place. On combustion, carbon monoxide liberates heat amounting to 68 Cals. per gram-molecule. The formation of carbon monoxide from carbon is also exothermic (*i.e.*, heat-producing), the heat liberated being 29 Cals. per gram-molecule. The following thermic equations may, therefore, be written—

$$2C + O_2 = 2CO + 58 \text{ Cals.}$$
$$2CO + O_2 = 2CO_2 + 136 \text{ Cals.}$$

and for complete combustion—

$$C + O_2 = CO_2 + 97 \text{ Cals.} \left(= \frac{194}{2}\right).$$

It is obvious, therefore, that in the conversion of carbon into carbon monoxide about 30 per cent. of the heat obtainable by complete combustion is evolved. Thus, for 1 lb. of carbon the theoretical yields are as under—

	C →	CO	→ CO_2.
Weight,	1 lb. →	$2\frac{1}{3}$ lbs.	→ $3\frac{2}{3}$ lbs.
Heat,	→	1,096 Cals.	→ 2,571 Cals.
		Total	→ 3,667 Cals.

Carbon monoxide is a very poisonous gas, even the presence of very small quantities in the air giving rise to unpleasant and dangerous symptoms. These effects are due to the fact that the carbon monoxide combines with the hæmoglobin of the blood. Traces of the gas may be detected by shaking the suspected air (after freeing, if present, from ammonia and sulphuretted hydrogen) with a solution of palladium chloride. 0.05 per cent. of the gas is sufficient to produce a precipitate. Cases of carbon monoxide poisoning should be treated by removing the patient into warm, fresh air, application of warmth, artificial respiration, and the administration of oxygen. Medical aid should be called in at once.

AIR-GAS

The following general description of a "gas-producer" is quoted from **Bone**:— "A modern gas-producer consists of a cylindrical furnace, 6-12 ft. internal

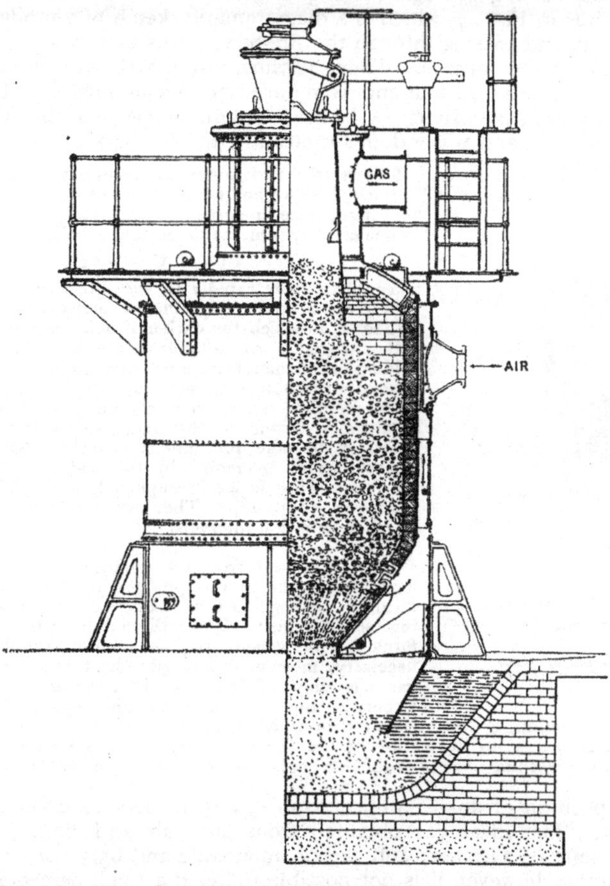

Fig. 26.—Water-Sealed Mond Producer. Showing how air is blown in through the circumferential grate at the bottom.

(Power-Gas Corporation, Ltd., Stockton-on-Tees.)

diameter, and 10-15 ft. high, lined with fire-brick with an outer mild steel casing, and usually water-sealed at the bottom. Means are provided (1) for the introduction of the blast through some form of grate or twyer fixed immediately above the level of the water in the seal, and its regular distribution through the bed of incandescent fuel (usually 3-5 ft. in thickness) which rests upon a lower bed of ashes; and (2) for the charging in of the raw fuel through (usually) some form of bell hopper fixed centrally on the top of the producer, round which are arranged a number of equidistant poking holes closed by iron balls. The gas outlet is fixed near the top of the cylindrical furnace."[1] (See Fig. 26.)

When such an apparatus having, however, a solid, not water-sealed bottom, is charged with any carbonaceous fuel (*e.g.*, coal or coke) and worked **with an air blast only**, the resulting gaseous fuel is known as "**air-gas**."[2] The amount of air supplied must not, of course, be sufficient to oxidise the fuel to the fullest extent, since what is aimed at is the production of the maximum amount of carbon monoxide and the minimum amount of dioxide. The air, meeting with the hot fuel at the bottom of the grate, may be considered to oxidise it mainly to carbon dioxide, but this, passing through the higher layers of hot fuel, is reduced to the monoxide. Thus in Fig. 27, which is a diagrammatic sketch of a producer, the hot air, entering at E, and passing through the fire-bars, burns mainly to carbon dioxide in the lower part of the furnace. Then, passing through the upper layers of red-hot fuel, it takes up more carbon and is reduced to the monoxide. The resulting air-gas passes through the exit C. A is the charging-hopper for the fuel, S S S are holes for poking, etc., and D is a door for cleaning the fire-bars.

From what has been said above, it follows that two things are necessary to the production of a good air-gas:—(1) The fuel must be stacked high; and (2) the temperature of the fuel must be kept high, since, in agreement with the results of **Rhead** and **Wheeler's** experiments already quoted, the reversible reaction,

$$CO_2 + C \longrightarrow 2CO,$$

takes place more rapidly in the direction of the arrow at high temperatures. Experiments carried out by **Rhead** and **Wheeler**, in which the carbon dioxide employed was diluted with nitrogen so as to approximate to the actual conditions holding in the manufacture of air-gas, show that "a temperature of over 1,100° C. must prevail in the fuel-bed if the fairly complete reduction of carbon dioxide desired is to be obtained." Owing to the high temperature needed, it is necessary to take precautions against clinkering troubles. These may be overcome by the addition of limestone to the fuel. This device is employed in the **Thwaite** cupola producer[3] (Fig. 28). The limestone forms a liquid slag which is tapped at intervals.

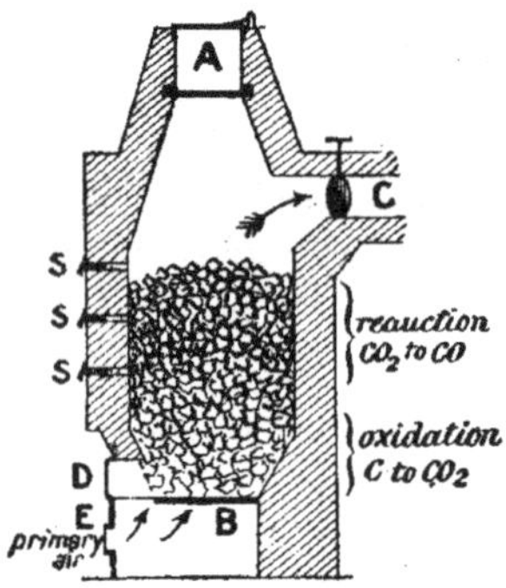

FIG. 27.—Section through a Producer.

(*With permission of Messrs Longmans, Green & Co.*)

It will be seen from the equations given under "carbon monoxide" that in the production of air-gas, nearly one-third of the total heat available from the fuel is converted into sensible heat. When the gas is to be burned in large furnaces, for which purpose preliminary cleaning is not necessary, most of this sensible heat may be utilised if the gas has only to travel a short distance to the furnace. Otherwise the sensible heat of the gas may be employed, by a system of regeneration, to heat the air with which it is to be burned; or it may be used for other purposes, according to the nature of the industry using the gas plant—the effecting of evaporation and the working of sulphur dioxide engines may be mentioned as two cases in point.

The composition of air-gas will, of course, largely depend upon the nature of the fuel used. The ideal gas obtained from pure air and pure carbon would consist by volume of 34.7 per cent. carbon monoxide and 65.3 per cent. nitrogen.[4] In actual practice, however, it is not possible to avoid a small percentage of carbon dioxide, and when coal (not coke) is used as the solid fuel, a small amount of methane (which is, however, valuable as a fuel) is always present owing to the distillation of the coal of the top layers in the producer.

The following may be given as two typical analyses. No. 1 refers to air-gas made in a

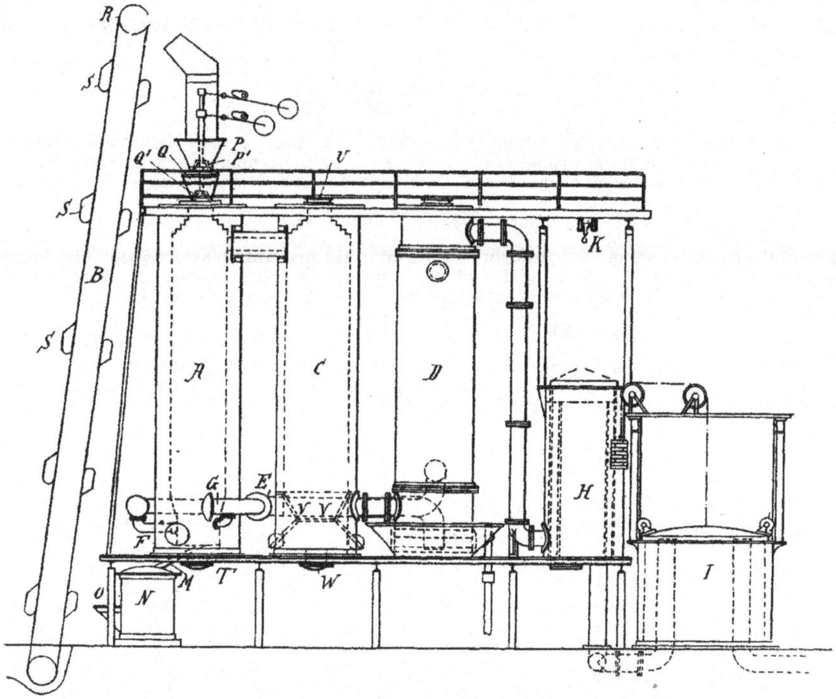

FIG. 28.—Thwaite Cupola Producer.

Thwaite producer from Lancashire slack coal, and is given by **Bone**. No. 2 refers to air-gas made from coke in a small **Dowson** plant, and is from **Dowson** and **Larter's** work referred to under "Literature."

	No. 1.		No. 2.	
	Per Cent.		Per Cent.	
Carbon monoxide - - - - - -	29.0		32.6	
Hydrogen - - - - - - -	5.35		1.0	
Methane - - - - - - -	2.05		...	
Total combustible gases - - -	——	36.4	——	33.6
Carbon dioxide - - - - - -	2.0		1.4	
Nitrogen - - - - - - -	61.6		65.0	
Total incombustible gases - -	——	63.6	——	66.4
Calorific power, B.T.U. per cubic foot -	130.8-138.7[1]		114.5-115.0	

WATER-GAS

In the manufacture of water-gas, the oxygen needed for the combustion of

[1] These are what are termed respectively the "net" and "gross" calorific powers. The "gross" value gives the amount of heat obtained when the products of combustion are cooled, so that liquid water is obtained; the "net" value gives the amount of heat obtained when (as in most cases of actual usage) the water is reckoned to be present wholly as steam. The percentages in these and following analyses refer to the volumes of the gases.

the solid fuel is obtained, not from the air, **but from steam.** Steam reacts with carbon according to one or other of the following equations :—

$$C + H_2O = H_2 + CO.$$
$$C + 2H_2O = 2H_2 + CO_2.$$

It is a matter of uncertainty which reaction takes place first: perhaps both carbon monoxide and carbon dioxide are formed simultaneously, as in the simple combustion of carbon. But in any case a final condition of equilibrium is reached through the reversible reaction :—

$$CO + H_2O \rightleftarrows CO_2 + H_2.$$

Both of the former reactions are strongly endothermic: the first absorbing about 29 Cals. per gram-molecule of carbon, whilst the second requires about 19 Cals. It is obvious, therefore, that these reactions will only take place when heat is supplied from without, and since it follows that

$$CO_2 + H_2 = CO + H_2O^1 - 10 \text{ Cals.},$$

it is obvious that high temperatures will favour the occurrence of the first reaction. These conclusions are borne out by experiment. The following results, due to **Dr Bunte,** are quoted from **Robinson's** work referred to under "Literature" (p. 547) :—

TABLE SHOWING THE ACTION OF STEAM ON CARBON (BUNTE).

Temperature in Degrees C.	Per Cent. of Steam Decomposed.	Composition, by Volume, of Gas Produced.		
		H_2.	CO.	CO_2.
674	8.8	65.2	4.9	29.8
758	25.3	65.2	7.8	27.0
838	41.0	61.9	15.1	22.9
954	70.2	53.3	39.3	6.8
1,010	94.0	48.8	49.7	1.5
1,060	93.0	50.7	48.0	1.3
1,125	99.4	50.9	48.5	0.6

Hahn has more recently studied the reversible reaction,

$$CO + H_2O \leftrightarrows CO_2 + H_2 + 10 \text{ Cals.},$$

and has found, in agreement with the law of mass-action, that the ratio, $\frac{CO \times H_2O}{CO_2 \times H_2}$ (where the symbols stand for the concentration of the gases present), is constant for any given temperature; and further, in accordance with the above conclusion, that the value of the constant increases with a rise in the temperature, from 0.81 at 786° C. to 2.49 at 1,405° C.

In order, therefore, to prevent the formation of more than a negligible proportion of carbon dioxide in the manufacture of water-gas, the reacting materials should be maintained at a temperature not below 1,000° C. In actual practice it has been found altogether unsatisfactory to supply the heat necessary to maintain this temperature by any system of external heating, owing to the loss of heat in transmission from without to within the producer. This method was at first attempted, but now the necessary heat is invariably obtained by combustion in air of part of the solid fuel employed. The process is intermittent; and in the earlier forms of it air-gas is produced alternatively with water-gas in the following manner :—The producer must be supplied with two outlets. First of all air is blown in, which produces air-gas and heats the fuel to the requisite temperature. This is known as the "hot-blow," "air-blow," or simply as the "blow," and lasts about ten minutes. The air-gas is taken off from one outlet, and is used to produce steam for the "cold-blow," "steam-blow," or "run," lasting about four minutes. Water-gas is meanwhile

[1] In this, as in all other thermo-chemical equations which follow in which water figures, the water is assumed to be in the state of vapour.

taken off at the other outlet. When the temperature has fallen below that at which the reaction can be advantageously carried out, the steam is cut off and air is again blown in to heat the fuel. However, the production of air-gas during the "blow" has now been given up, in preference to plants in which the whole of the heat obtainable from a portion of the fuel is used to produce the desired temperature. This has led to a great reduction in the time of the "blow."

In Messrs **Dellwik** and **Fleischer's** water-gas producer,[1] the layer of fuel is less deep than in the older forms. The object of this is that during the "blow" that portion of the fuel consumed is burnt completely to carbon dioxide. By this means the requisite temperature is obtained far more quickly—the blow lasts only about one minute—and at the expense of less fuel. The apparatus is shown diagrammatically in Fig. 29. Of course, when carbon dioxide instead of air-gas results from

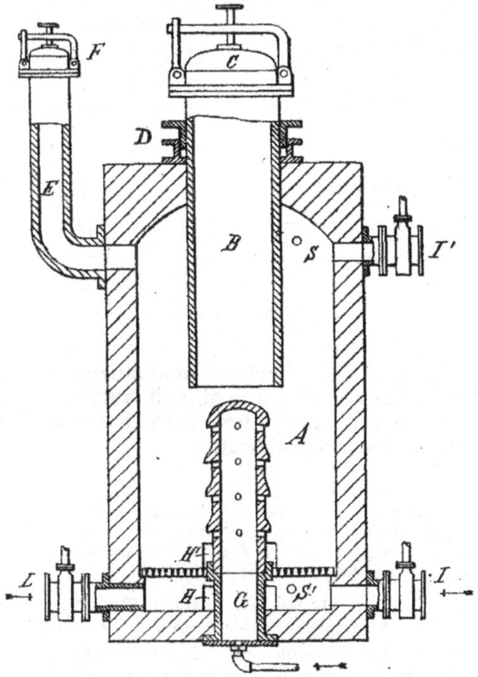

FIG. 29.—Dellwik and Fleischer's Plant.

the "blow," extra fuel must be burnt to produce the steam required for the "run," but it is found that there is a considerable saving in fuel on the whole, as well as a considerable saving of time.

In order to render the production of gas continuous, plants have been devised in which two producers are worked together, one being "blown," whilst the other is "run," and *vice versa.*

"A is the generator provided with a coal receptacle B . . . of cast iron; this coal receptacle passes through a stuffing box D placed on the cover or top of the generator. Under the grating is arranged the main air inlet L, and the gas outlet I, and steam inlet S'. The generator is also provided with cleaning doors H and H'. In the upper part there is an outlet pipe E, for the combustion products, which pipe is closed during the period of the production of water-gas by the valve or cover F. I' is an outlet for water-gas and S is a second steam inlet. The coal receptacle B is charged with coal, coke, or other suitable fuel, and then closed by the door or cover C. The air inlet is arranged under the grate and preferably continued into the layers of fuel by an appropriate pipe G, provided with numerous air nozzles or apertures, so that a thorough contact of the air with the whole mass of fuel is obtained. When the proper height of a given description of fuel for a given amount of air pressure has been determined by testing by means of the adjustable receptacle B shown in the figure, the cast-iron receptacle is advantageously replaced by one made of fire-bricks."

[1] English Patent, 29,863, 1896.

Ideal water-gas should consist of 50 per cent. of carbon monoxide and 50 per cent. of hydrogen by volume. But in actual practice, of course, small amounts of carbon dioxide and nitrogen cannot be avoided.

The following may be given as typical analyses :—No. 1 refers to water-gas made at Essen from coke by the older process in which air-gas is produced during the "blow," and is taken from **Fischer's** "Handbuch der Chemischen Technologie" (15th ed.), Vol. I., p. 50. No. 2 refers to water-gas made in the improved plant of **Dellwik** and **Fleischer**. The two samples differ in composition very slightly from one another.

	No. 1.		No. 2.	
	Per Cent.		Per Cent.	
Carbon monoxide	44.0		43.0	
Hydrogen	48.6		49.0	
Methane	0.4		0.5	
Total combustible gases	——	93.0	——	92.5
Carbon dioxide	3.3		4.0	
Nitrogen	3.7		3.5	
Total incombustible gases	——	7.0	——	7.5
Calorific power, B.T.U. per cubic foot	295.5-323.5		290-340	

SEMI-WATER-GAS

Semi-water-gas is produced by blowing simultaneously steam and air through red-hot coke or other carbonaceous material, and is by far the most important type of producer-gas, so much so that it is frequently meant when the term "producer-gas" is employed without further qualification.

It was pointed out under "Air-Gas" that in manufacturing this from carbonaceous fuel, about 30 per cent. of the available heat was converted into sensible heat. This, of course, is a source of waste, though, as already pointed out, a good deal of this heat may be utilised by means of various devices. Now if steam is passed through the fuel in the producer as well, it is obvious that one or other of the endothermic reactions between it and carbon,

$$C + H_2O = CO + H_2 - 29 \text{ Cals.}$$
$$C + 2H_2O = CO_2 + 2H_2 - 19 \text{ Cals.},$$

will take place at the expense of this sensible heat. In other words, one or other of these reactions will convert the sensible heat into potential chemical energy, which will be obtainable again as sensible heat by combustion of the resulting gaseous fuel. The exact order in which the various reactions between the carbon, oxygen, and water take place is rather uncertain, but in any case a condition of equilibrium is obtained through the two reversible reactions dealt with under "Carbon Monoxide" and "Water-Gas" respectively, namely—

$$CO_2 + C \rightleftarrows 2CO - 39 \text{ Cals.},$$
and
$$CO + H_2O \rightleftarrows CO_2 + H_2 + 10 \text{ Cals.}$$

And the result is the same as if one or other of the above reactions took place between carbon and steam concomitantly with the combustion of another portion of the carbon in the air to carbon monoxide, *i.e.*,

$$2C + O_2 = 2CO + 58 \text{ Cals.}$$

By adjusting the ratio between the amounts of steam and air, it ought to be possible to absorb all the heat given out by this last reaction, by that between the steam and the carbon. And for this purpose the two forms of this reaction are almost as good as one another, since that in which carbon dioxide is produced gives an additional amount of hydrogen, with a heat of combustion but little less than that of the carbon monoxide otherwise formed (58 Cals. net per gram-molecule in the first case, to 68 Cals. in the second), though the formation of carbon dioxide serves to dilute the gas somewhat, giving a larger volume of gas, but decreasing the calorific power of a given volume. But, of course, this is not actually possible, since there is a temperature below which the reactions will not take place at all, and to maintain this it is necessary that the sum total of the reactions should be exothermic. Moreover, in actual practice an additional amount of carbon dioxide is always produced. But an approximation to the ideal condition of things is made, and the addition of the steam blast to the air blast increases the economy of production enormously, and yields a gas at a less high temperature than with air blast alone, and of a considerably increased calorific power. When it is necessary to cool the gas (*e.g.*, for use in gas engines), its sensible heat may be used, by a system of regeneration, to heat the air-blast.

The extent to which the reactions,

$$C + H_2O = CO + H_2,$$
and
$$C + 2H_2O = CO_2 + 2H_2,$$

take place depends upon the ratio of the steam blast to the air blast. For furnace purposes the hydrogen content of the gas should not exceed 14 per cent., hence it is desirable that the first and not the second of these reactions should occur. On the other hand, this is not of importance if the gas is to be employed for power purposes; and by the use of an excess of steam, in which case the second rather than the first of these reactions takes place, it is possible to recover, in the manner described below, much of the nitrogen of the fuel as ammonia.

Ideal producer-gas made by the reactions—

$$2C + O_2 = 2CO,$$
and
$$C + H_2O^{1} = CO + H_2,$$

would have a composition by volume of CO, 39.9 per cent.; H_2, 17.0 per cent.; and N_2, 43.1 per cent.; whereas that made by the reactions—

$$2C + O_2 = 2CO,$$
and
$$C + 2H_2O^{1} = CO_2 + 2H_2,$$

would have a composition by volume of CO, 19.8 per cent.; H_2, 28.7 per cent.; CO_2, 14.4 per cent.; and N_2, 37.1 per cent.

The following typical analyses by **Sexton** of various samples of producer-gas show an approximation to the first of these ideal gases, except No. 4, which approximates to that of the second type; but in every case, as might be expected, the nitrogen content is much in excess of the theoretical.

Analysis of Semi-Water-Gas (Sexton)

	No. 1.	No. 2.	No. 3.	No. 4.	No. 5.
	Per Cent.	Per Cent.	Per Cent.	Per Cent.	Per Cent.
Carbon monoxide	24.40	26.40	27.00	16.15	20.40
Hydrogen	8.60	12.13	10.90	19.43	12.60
Methane, etc.	2.40	2.00	1.28	2.66	3.50
Carbon dioxide	5.20	9.16	4.50	11.53	5.50
Nitrogen	59.40	50.31	56.32	50.23	58.00

In 1906-7 **Bone** and **Wheeler** (*loc. cit.*) carried out a series of experiments to determine the influence of varying blast steam saturation temperatures upon the composition and yield of the gas from a Lancashire bituminous coal, the results of which are tabulated on p. 56.

[1] Steam raised by part of the heat of the accompanying exothermic reaction.

RESULTS OF EXPERIMENTS BY BONE AND WHEELER.

Average depth of incandescent fuel	$3\frac{1}{2}$ ft.					7 ft.				
Average rate of gasification (day-shift) per hour per producer	22.5 cwt.					11.5 cwt.				
Steam saturation temperature of blast	45°	50°	55°	60°	70°	60°	65°	70°	75°	80°
Composition of gas, per cent.—										
Carbon monoxide	31.60	30.60	28.10	27.30	20.85	27.30	25.40	21.70	18.35	16.05
Hydrogen	11.60	12.35	15.45	15.50	19.75	16.60	18.30	19.65	21.80	22.65
Methane	3.05	3.00	3.00	3.05	3.45	3.35	3.40	3.40	3.35	3.50
Carbon dioxide	2.35	2.50	4.40	5.10	9.25	5.25	6.95	9.15	11.65	13.25
Nitrogen	51.40	51.55	49.05	49.05	46.70	47.30	45.90	46.10	44.85	44.55
Calorific power in B.T.U. per cubic foot at 0° and 760 mm. — Gross	180.0	178.5	180.8	178.7	175.5	185.6	185.4	177.5	172.0	169.5
Calorific power in B.T.U. per cubic foot at 0° and 760 mm. — Net	170.5	168.7	169.1	166.9	161.0	173.0	172.0	163.3	157.3	154.3
Yield of gas, cubic feet at 0° and 760 mm. per ton coal	133,700	132,500	132,700	135,000	...	138,250	134,400	141,450	145,800	147,500
Steam added to blast, pounds per pound coal	0.2	0.21	0.32	0.45	...	0.45	0.55	0.80	1.10	1.55
Per cent. steam decomposed	All	All	All	76.0	...	87.0	80.0	61.0	52.0	40.0

TYPES OF GAS-PRODUCERS

It will be impossible here to attempt to give a description of the very many different kinds of gas-producers now in use. A great deal of attention has recently been devoted to the subject, and very many patents have been taken out in recent years, though a large proportion of these are concerned only with improvements in matters of detail. For consideration of such matters the reader should consult the special books dealing exclusively with the subject noted under "Literature." All that can be done here is to give a brief description of the most important and typical producers.

The simple form of gas-producer with a flat grate near the bottom is now practically obsolete, owing to the difficulty of removing the clinker when such a grate is employed. The chief aim of modern producers is to effect the various stages of gas-production by mechanical means as far as possible, and thus render the whole process continuous. Modern gas-producers may be roughly classified according to the type of grate used and the manner in which the air blast is admitted, as under:—

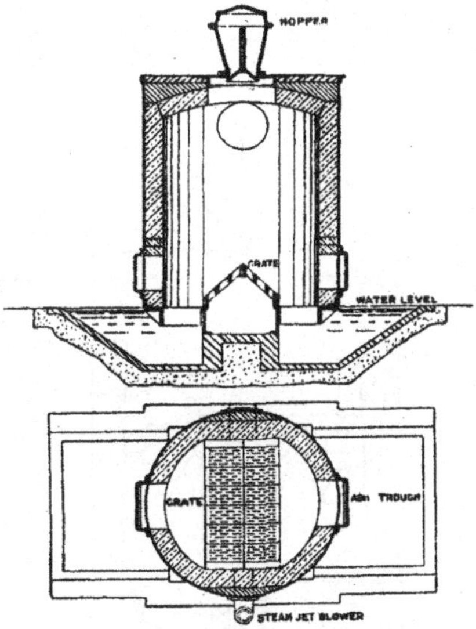

FIG. 30.—The Duff Producer (two views).

(*Taken from Mr F. G. Fogg's article in "The Chemical World" of Nov. 1912, and reproduced with kind permission of the editor, Mr Dreaper, and the publishers, Messrs J. & A. Churchill.*)

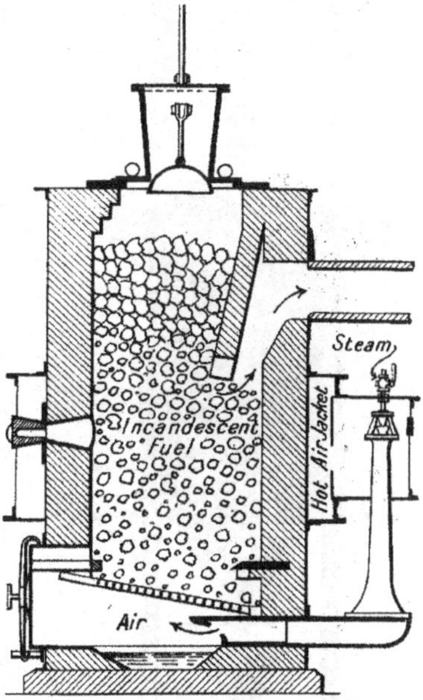

FIG. 31.—Thwaite "Simplex" Producer.

1. **Diagonal Grate Producers.**—In the **Duff** producer (Fig. 30) the grate-bars, as will be noticed from the diagram, run across the bottom of the producer, not, however, occupying the whole area, forming in section an inverted "V." The air blast enters beneath the grate, whose form ensures that the air is uniformly distributed over a large area of fuel, and readily admits of the clinker being pushed into the water trough below. In the **Thwaite** "Simplex" producer (Fig. 31) the grate, underneath which the air blast and steam enter, slopes in one direction only. This producer is also water-sealed.

2. **Circumferential Grate Producers.**—The **Mond** producer belongs to this type, the grate being made of an inverted truncated cone of fire-bars through which the blast enters. The shape admits of the ready removal of the ash. The

producer is made both with a water-seal for fuels giving a small percentage of ash (Fig. 26) and with a mechanical dry ash discharge, which automatically delivers the ash from the bottom of the apparatus for use with fuels having a high ash content (Fig. 32). For the special advantages of the **Mond** apparatus see below, "Ammonia Recovery in the Manufacture of Semi-Water-Gas."

3. **Central Blast Producers.**—In producers of this type the blast enters through a louvred cone (*i.e.*, a cone consisting of a number of plates) placed centrally near the bottom of the producer, the grate-bars being usually dispensed

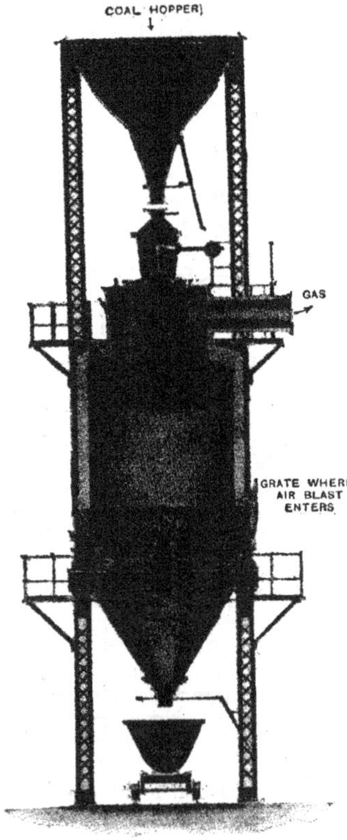

FIG. 32.—The Mond Producer, with Mechanical Dry Ash Discharge.

(*The Power-Gas Corporation, Ltd., Stockton-on-Tees.*)

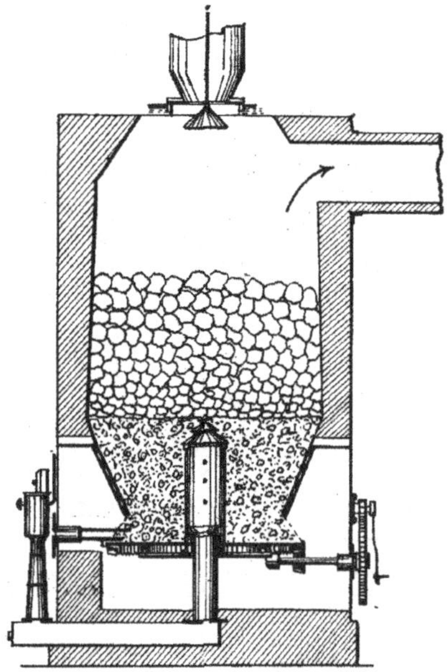

FIG. 33.—The Taylor Revolving Bottom Gas Producer (internal diameter, 6 feet).

with. The **Taylor** "Revolving Bottom," **Talbot** and **Kerpeley** producers may be instanced as well-known examples of this type. In the **Taylor** producer (Fig. 33) the mixture of steam and air is delivered by means of a pipe, fitted with a conical distributor, rather high up in the apparatus. The fuel and clinker rest upon a revolving plate, underneath which is a closed ash-pit. By rotating the bottom when necessary, the height of the clinker is maintained at about six inches above the level of the steam and air inlet, and, by this means also, air-passages that may have been formed by the action of the blast are closed up. The **Talbot** producer (Fig. 34) is provided with a central blast of the usual type, and a double mechanical feed, as shown in the diagram. The apparatus is water-sealed and the clinker is removed by hand from the water trough. The **Kerpeley** producer (Fig. 35) has a revolving grate and hearth. The blast distributor, which is fixed to the grate, is in the shape of a cone with its apex rounded. The lower portion of the apparatus is cooled by means of a water jacket. This obviates the clinker adhering to the sides, which would interfere with the rotation of the grate. The ashes in the rotating water trough are deflected over the side by means of a

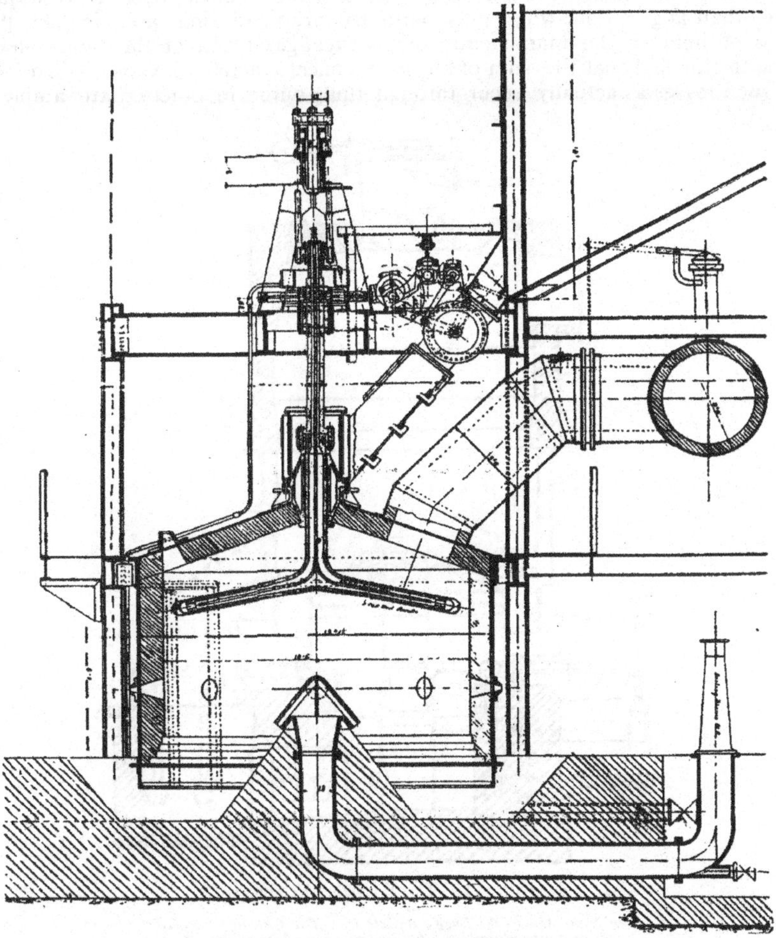

FIG. 34.—The Talbot Producer.

(Taken from Mr F. G. Fogg's article in "The Chemical World," Dec. 1912, and reproduced with kind permission of the editor, Mr Dreaper, and the publishers, Messrs J. & A. Churchill.)

fixed blade set at a convenient angle. (For a description of suction gas-producers see p. 63.)

USES AND ADVANTAGES OF PRODUCER-GAS

Semi-water-gas is employed both in furnaces and in gas engines. For the former purpose, when the furnace is a large one, the gas need be neither cooled nor scrubbed clean. In the case of small furnaces, where there are small stop-cocks, etc., it is desirable, to avoid clogging, to scrub the gas free from tarry matter; and for use in gas engines the gas must be both clean and cool. Many improvements have been effected in gas-fired furnaces since **Siemens** patented his regenerative furnace in 1861; and great advances have been made in the construction of gas engines in recent times. Such subjects, however, lie without the domain of this work.

The advantages of producer-gas over the direct consumption of solid fuel are numerous. Of course, it is not possible to obtain more heat from a given quantity of fuel than is given out when it is burnt to carbon dioxide, and, in fact, there is a loss of heat in the manufacture of producer-gas (even of the semi-water type) owing to the fact that the sum of all the chemical reactions involved is exothermic. But such losses as actually occur through this source in practice are amply made

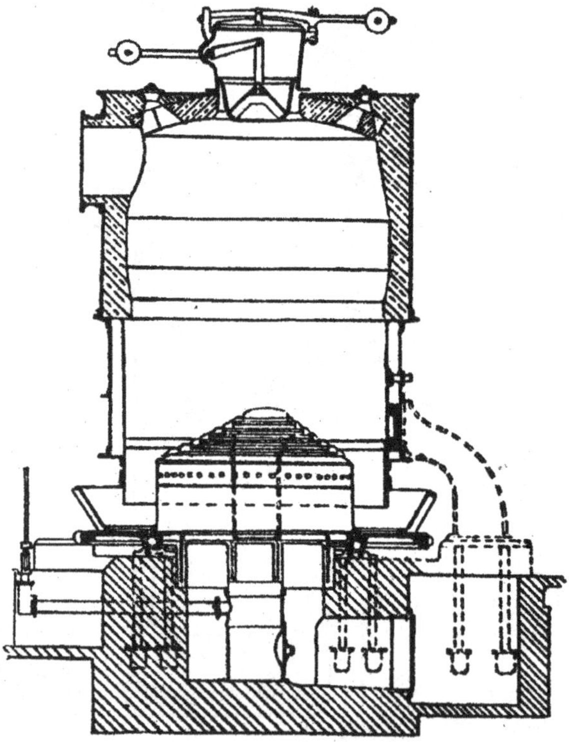

FIG. 35.—The Kerpeley Producer.

(*Taken from Mr F. G. Fogg's article in "The Chemical World," Dec. 1912, and reproduced with kind permission of the editor, Mr Dreaper, and the publishers, Messrs J. & A. Churchill.*)

up for by the economic manner in which gas can be burned as compared with solid fuel. There are no losses in the case of gas-furnaces through smoke or convection currents. The amount of air required for combustion is not much above the theoretical, whereas in furnaces burning solid fuel far more air is necessary than actually takes part in the combustion, so that there is in the latter an extra large volume of gas to carry away sensible heat through the chimneys. Moreover, in gas-fired furnaces the sensible heat of the escaping gases is not wasted but is used on the "regenerative" principle to heat the air for combustion, and **Bone** and **M'Court's** method of surface combustion promises still further advantages in the economic combustion of gaseous fuels.

Furthermore, gas is more easily transmitted than solid fuel, gas furnaces may be maintained at a constant temperature, and a gas fire may be more easily applied in metallurgical work than one burning solid fuel. It is important to notice, also, in connection with metallurgical work, that with gas an oxidising or reducing flame may be obtained at pleasure.

Nothing can be urged against the use of producer-gas as compared with coal on the grounds of efficiency or economy. One objection is that the gas is very poisonous, and that deadly quantities of it may escape into the air without betraying their presence by their odour. But so long as due care is exercised, especially with suction plants to be described later, nothing need be feared on this score. One way out of the difficulty is to flavour the gas, as is done in Germany in the case of water-gas, by bubbling it through mercaptan, a compound which has a pungent and most disagreeable odour.

Something in favour of the use of producer-gas as against solid fuel may be said from another point of view. There can be no doubt that the smoke nuisance is one of the evils of modern times, and renders every great manufacturing town unsightly and unhealthy. **Sir Oliver Lodge**,[1] as well as other writers, have written eloquently on this subject, and dream of the days when coal will invariably be converted into either coal-gas or producer-gas at the pit's mouth, and conveyed to consumers in the form of a clean, non-smoke-producing fuel. Little has been done as yet to supply producer-gas in the way that coal-gas is supplied, it being usual for consumers to make their own gas. But in February 1905 an installation was built by **The Power Gas Corporation, Ltd**. (the holders of the **Mond** patents), for **The South Staffordshire Mond Gas Co**, formed for the purpose of manufacturing and distributing producer-gas throughout a district of about 120 square miles. The plant has run continuously day and night, distributing gas in steel pipes to a large number of consumers using the gas for many different purposes—power, annealing, enamelling, japanning, oxide burning for paint, heating, galvanising, etc.

Finally, as will appear below, producer-gas may with advantage be produced from fuels of too poor a quality to be worth utilising otherwise.

The above remarks apply mainly to gas of the semi-water type. Air-gas is by no means so economical, and is used far less extensively, though some engineers prefer it owing to its freedom from hydrogen, for reasons which seem by no means obvious.

The uses of water-gas depend to some extent upon its greater calorific power as compared with other types of producer-gas, which greater calorific power is due to the fact, of course, that water-gas contains only small quantities of nitrogen, whereas nitrogen is present in large quantities in air-gas and semi-water-gas. It is used in Germany, on this account, for steel-plate welding. Another use for the gas is found in the fact that, although when pure it burns with a pale blue, non-luminous flame, after being carburetted with the decomposition products of petroleum, it gives a satisfactory luminous flame. Such gas is called "carburetted water-gas," whilst the non-carburetted variety is distinguished as "blue water-gas." Water-gas is also used as an illuminant by systems in which "mantles" (*e.g.*, the **Welsbach** mantle) or rods (**Fahnehjelm** system) of refractory material are heated by the combustion of the "blue" non-luminous gas to such a temperature that they emit a strong white light. (For further details see **Mr Dancaster's** article on "Illuminating Gases" in **Martin's** "Industrial Chemistry," Vol. I.) More recently water-gas has become very important as a source for the manufacture of hydrogen. (See the present writer's article on "Hydrogen" in this volume.)

AMMONIA RECOVERY IN THE MANUFACTURE OF SEMI-WATER-GAS

As mentioned in the section on "Semi-Water-Gas," it is possible, by means of certain modifications in the method, to recover the nitrogen of nitrogen-containing fuels as ammonia. The **Mond** system was specially devised for this end, though **Mond** plants may also be used for manufacturing producer-gas without nitrogen recovery. It is claimed that by the **Mond** process as much as 70 per cent. of the nitrogen present in the solid fuel may be recovered, and as there is a large demand for ammonium salts as fertilisers, the process is exceedingly profitable in the case of large plants working with fuel of good nitrogen-content. Fig. 26, p. 49, shows the construction of the Mond producer for ammonia recovery.

The producer, as stated on p. 58, is water-sealed, and the grate is circumferential. The blast must contain a large excess of steam—1 lb. of bituminous fuel requires 3 lbs. of air and 2½ lbs. steam—which serves to keep the temperature down, and thus prevents the decomposition of the ammonia formed, whilst at the same time it minimises the formation of clinker, and so allows of the continuous working of the apparatus. Only about one-fifth part of the steam is actually decomposed, the rest escaping with the gas, part of the heat of which is recovered by means of a tubular regenerator, and is used to raise steam (about 40 per cent. of the total) for the blast. The blast is further heated by being circulated around the producer within an outer casing. The gas is further cooled down to about 90° C. by means of a water spray, which frees it from tar (which may be recovered from the washings). It then passes into a lead-lined tower, where it meets with a stream of ammonium sulphate solution containing about 4 per cent. of free sulphuric acid. The acid extracts the ammonia from the gas,

$$2NH_3 + H_2SO_4 = (NH_4)_2SO_4,$$

giving ammonium sulphate. The gas is finally washed with cold water before being used.

For further details regarding Ammonia Recovery, see **Martin's** "Industrial Chemistry," Vol. II.

Owing to the large proportion of steam used, the reaction between the steam and the carbon of the fuel takes the form—

$$C + 2H_2O = 2H_2 + CO_2.$$

Consequently the gas differs considerably in composition from ordinary semi-water-gas, containing more hydrogen and carbon dioxide and less carbon monoxide. Moreover, it has a smaller calorific power. But with very large plants working with highly nitrogenous fuels, the gas may be obtained practically without cost owing to the value of the ammonia recovered.

The process, moreover, can be applied with success to peat containing as much as 60 per cent. to 70 per cent. of water, and brown coal containing 50 per cent. to 60 per cent. of water, which cannot be gassified by other processes without being first dried—which cannot be economically effected. For smaller consumers **Mond** plants are made without ammonia recovery, in which case less steam is employed, and a gas containing less hydrogen and carbon dioxide, and more carbon monoxide is obtained.

The following analyses of "**Mond gas**" are published by **The Power Gas Corporation.**[1] No. 1 refers to bituminous fuel with ammonia recovery. No. 2 refers to bituminous fuel without ammonia recovery.

	No. 1.		No. 2.	
	Per Cent.		Per Cent.	
Carbon monoxide	11.0		23.0	
Hydrogen	27.5		17.0	
Methane	3.0		3.0	
Total combustible gases	—	41.5	—	43.0
Carbon dioxide	16.5		5.0	
Nitrogen (and moisture)	42.0		52.0	
Total incombustible gases	—	58.5	—	57.0
Calorific power, B.T.U. per cubic foot	147.4-164.3		157.5-169.2	

[1] To whom my thanks are due for kindly supplying me with pamphlets descriptive of the various types of **Mond** producers, etc.

SUCTION GAS-PRODUCERS

In ordinary producer-gas plants the gas is made under a pressure somewhat above the atmospheric, the blast being forced into the apparatus. For furnace purposes the gas may be taken direct from the producer, but when required for gas engines it must first be stored in a gasholder and then delivered under very slight pressure. To avoid the expense of a gasholder, producers have been devised in which the gas is produced moment by moment, as required, by the sucking action of the gas engine driven by the gas produced. Each movement of the piston draws a fresh supply of gas from the producer, and an equal volume of air and water-vapour is consequently sucked into it.

The generation of gas must be initiated by means of a blower, and the engine is not started until the gas is found to burn satisfactorily at the test-cock. The gas is cleaned in a scrubber, and preferably dried in a dryer before entering the engine. The production of gas continues so long as the engine is working, and ceases when the engine stops and, thus, no longer sucks air and water-vapour into the producer. Suction plants, of course, give most satisfactory results when the engine is running at full load, as otherwise the quality of the gas is liable to deteriorate.

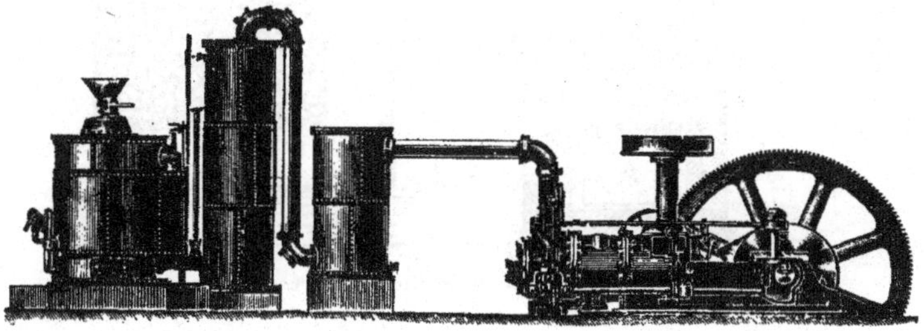

FIG. 36.—Mond Suction Plant and Engine.

(*Power-Gas Corporation, Ltd.*)

A further economy is effected by the fact that no boiler is required in conjunction with these plants, its place being taken by a "vaporiser" or "evaporator," of which perhaps the best form is that of an internal annular boiler, as in the **Mond** suction producer (Fig. 36), the Dowson plant, and others. This raises steam by means of the waste heat of the producer, the needed air also being sucked into the apparatus through the vaporiser.

The details of the Dowson (30 B.H.P.) plant, as a modern and fairly typical suction producer, are shown diagrammatically in Fig. 37, and the following description of the apparatus and the manner of working it is quoted from **Dowson** and **Larter's** work referred to under "Literature." "A is the grate on which the fuel rests; B is the container holding a store of fuel which is admitted through the hopper and valve at the top; C is a chamber filled with loose pieces of fire-brick; D is a circular pipe with holes in the top, which receives water from the feed-pot E; F is the air inlet, and G the gas outlet, with test-cock at H; I is a chimney or waste pipe; J is a coke scrubber with water-seal at the bottom; K is the gas outlet. . . . A small hand-power fan (not shown in the figure) [is used] for blowing up the fire at the start. When the plant is to be worked, a little oily waste and some wood are put on the grate A, and the producer is filled with anthracite or coke in small pieces. The feeding hopper is then closed, and the cock on the waste pipe I is opened; the fire is then lighted through a door on the level of grate A; the fan is set in motion, and the products are in the first instance allowed to escape through the pipe I. The water supply is turned on, and as soon as gas will burn well at the test-cock H, it is ready for use in the engine; the waste pipe I is then closed, and the gas is blown off through a waste pipe near the engine until good gas from the producer reaches the gas-cock on the engine. The engine is then started and the fan is stopped. From this time the engine itself draws in the air required for the producer, as well as the water to be vaporised in the chamber C, the mixture of steam and air passing downwards from C, through L, to the underside of the grate A."

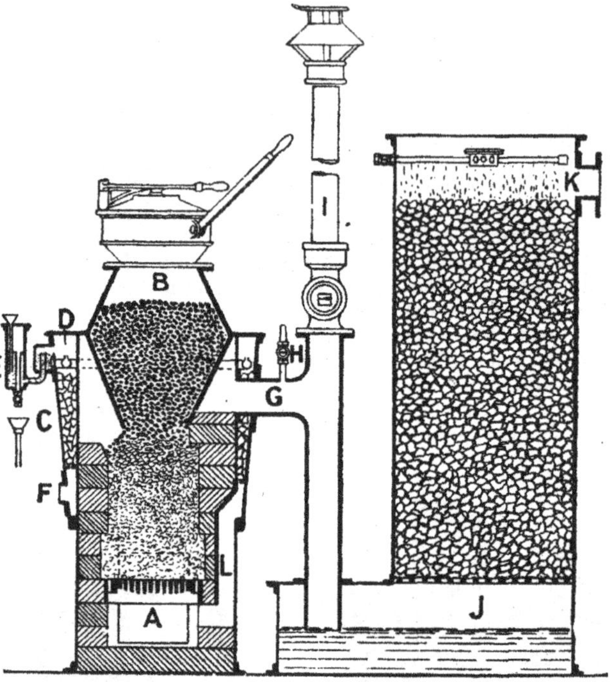

FIG. 37.—Dowson Suction Gas-Producer.

BLAST-FURNACE-GAS

The name "blast-furnace-gas" is given to the waste gas derived from furnaces used in the smelting of iron and other ores. In composition the gas is the same as a low grade producer-gas, its calorific power varying from about 90 to 100 B.T.U. per cubic foot. It is frequently employed as a gaseous fuel, and thus, as it resembles producer-gas in its composition and its industrial uses, it may be dealt with in the present place.

The composition of the gas is variable, but the following figures are given by **Blount** and **Bloxam** as typical analyses, on the authority of **Turner**.

In the case of No. 1 coke was the fuel employed, whilst bituminous coal was used in the case of No. 2:—

	No. 1.		No. 2.	
	Per Cent.		Per Cent.	
Carbon monoxide	25		28.0	
Hydrogen	2		5.5	
Methane	2		4.4	
Total combustible gases	—	29	—	37.9
Carbon dioxide	12		8.6	
Nitrogen	59		53.5	
Total incombustible gases	—	71	—	62.1

In some cases, however, the content of carbon monoxide may be as high as 30 per cent.

Until 1837, when attention was called to the matter by **Dufaur,** this valuable source of energy had been entirely neglected. Plants were then devised and erected whereby about 50 per cent. of the blast gases were employed to heat the blast, being burnt in furnaces of the "regenerative" type, and about 40 per cent. were used for boiler firing to produce steam, the remaining 10 per cent. being wasted. But the efficiency of such boilers was unsatisfactorily low.

Nowadays the improvements that have been effected in gas engines and the introduction of large gas engines have rendered possible the more efficient utilisation of blast-furnace-gas, and it is now extensively used for this purpose. Before entering the engine the gas must be cooled to about 19° C., and the dust, which, of course, is usually present in such gas in considerable quantities, must be reduced to not more than 0.01 g. per cubic metre—a quantity which is negligible. To effect this the gas is first passed through one or other of the usual forms of dry-cleaning apparatus. It is then washed with water, a centrifugal apparatus being employed, either alone or after ordinary water-washing, in which the gas is thoroughly agitated with water in the form of fine spray. Ammonia, as well as tar, can be recovered from the water used for washing gases from blast-furnaces in which raw coal is burnt.

CHAPTER VI

The Carbon Dioxide (Carbonic Acid) Industry

CHAPTER VI

THE CARBON DIOXIDE (CARBONIC ACID) INDUSTRY

By H. Stanley Redgrove, B.Sc.(Lond.), F.C.S.

OCCURRENCE

Carbon dioxide is invariably present in the atmosphere, being formed by the respiration of animals as well as by the decay of organic matter. The amount varies from about 0.03 per cent. in "fresh" air to as much as 0.3 per cent. in crowded rooms. For the same reason it is invariably to be found in the soil, from which it is taken up by rain and spring water.

Enormous quantities of carbon dioxide are evolved from volcanoes, and from fumaroles in volcanic districts, the Grotto del Cane near Naples being specially noted in this respect. Large amounts are also present in many natural mineral waters, especially those coming from springs of volcanic origin, as in Germany, Java, etc.

The bulk of the world's content of carbon dioxide occurs stored up in rocks and minerals, such as chalk, limestone, marble ($CaCO_3$), magnesite ($MgCO_3$), dolomite ($MgCO_3.CaCO_3$), etc.

PROPERTIES

Carbon dioxide is a binary compound of the formula CO_2, and having a molecular weight of 44 (O = 16). It contains 27.27 per cent. of carbon, and 72.73 per cent. of oxygen by weight. It is a colourless gas, possessing a slightly pungent smell and an acid taste. It is very heavy, and can readily be poured from one vessel to another, its specific gravity at N.T.P. being 1.529 (air = 1, **Rayleigh** and **Leduc** independently), so that 1 litre of the gas at N.T.P. weighs 1.977 g.

The gas is non-combustible, and will not support the combustion of other bodies, with certain exceptions, such as magnesium, which will burn in carbon dioxide, yielding magnesium oxide and free carbon. The presence of 4 per cent. of carbon dioxide in air is sufficient to extinguish a burning candle. It is not poisonous, but it will not support respiration, so that air containing much carbon dioxide asphyxiates; hence asphyxiating air is often found in wells, cellars, mine-shafts, etc.

Carbon dioxide is sometimes known as carbonic acid gas. It is, correctly speaking, the anhydride of carbonic acid (H_2CO_3), which acid is formed when carbon dioxide is dissolved in water ($H_2O + CO_2 = H_2CO_3$). This is shown by the fact that the solution exhibits a weakly acid reaction to litmus, which reaction is given by neither the dry gas nor by dry liquid carbon dioxide (*q.v.*). This acid forms two series of salts, the carbonates (M_2CO_3) and the bicarbonates ($MHCO_3$) (where M is any univalent metal), which evolve carbon dioxide when acted on by acids less volatile than that gas.

The solubility of carbon dioxide in water is high, the "absorption coefficient" (the volume of gas measured at 0° C. and 760 mm. which is absorbed by 1 c.c. of water at $t°$ C. and 760 mm.) being $1.7967 - 0.07761t + 0.0016424t^2$, so that at ordinary temperatures the gas is soluble in about its own weight of water. The solubility for a few temperatures is given in the table below:—

Temperature, Degrees C.	0.	5.	10.	15.	20.
Solubility (volume CO_2 at N.T.P. absorbed by 1 c.c. water at 760 mm.)	1.7967	1.4497	1.1847	1.0020	0.9014

The solubility is increased by pressure, but not quite to the extent demanded by **Henry's** law (according to which the ratio between the solubility and the pressure should be constant). According to **Wroblewski** (Wiedemann's "Annalen der Physik und Chemie," Vol. XVIII., 1883, p. 290), the solubilities at 12.4° C. for various pressures are as under:—

Pressure in atmospheres - -	1	5	10	15	20	25	30
Solubility at 12.5° C. - -	1.086	5.15	9.65	13.63	17.11	20.31	23.25

On removing the pressure, and especially in the presence of a porous body like bread or sugar, the additional volume of the gas is briskly evolved, causing effervescence, as in the case of soda water, beer, etc. The solution has a sharp, refreshing taste.

Liquid and solid carbon dioxide will be dealt with below.

MANUFACTURE, ETC.

1. **Collection and Purification of Natural Carbon Dioxide.**—In order to collect the carbon dioxide escaping from mineral springs, borings are made in suitable regions, and round each boring a basin is excavated. The basin is kept filled with the water escaping from the bore-hole, the excess of water running away by means of an overflow channel. Over the bore-hole is placed a metallic gas-collecting vessel, or gasometer, which is provided with a counterpoise. The escaping gas collects under the gasometer, and is led away by a tube passing out of the upper part. The gasometer is automatically raised or lowered in the water by the counterpoise, according as the quantity of gas collecting in it increases or decreases. A similar apparatus is in use for collecting the carbon dioxide evolved from dry earth fissures.

The escaping gas is cleaned in a spray apparatus, in which it is absorbed by water under pressure in the presence of coke. A small quantity of potassium permanganate may be added more effectively to oxidise such impurities as sulphuretted hydrogen. The gas is then dried in towers containing calcium chloride, and is finally liquefied and stored in steel cylinders.

South Germany has many important carbon dioxide springs, from which enormous quantities of the gas are obtained. The Saratoga Springs, New York, U.S.A., must also be mentioned as of considerable importance.

2. **As a By-product in the Manufacture of Beer, Etc.**—Enormous quantities of carbon dioxide are evolved in the manufacture of beer, wines, spirits, etc., owing to the sugar undergoing alcoholic fermentation—

$$\underset{\text{(Sugar)}}{C_6H_{12}O_6} = \underset{\text{(Alcohol)}}{2C_2H_5OH} + 2CO_2.$$

The gas thus evolved has a flavour like fusel oil, and is, therefore, unsuited for many of the purposes for which carbon dioxide is in demand; but fermentation carbon dioxide is certainly suitable for carbonating and raising beer, and many methods have been devised for economically collecting it, which have, however, met with but little success. **Stead** and **Palmer** (English Patents, 27,396, 1904, and 23,763, 1907), for instance, collect the gas by means of counterbalanced gasometer bells, so constructed as to operate by their vertical movement means for throttling the flow of gas, effected by the use of a sleeve fitting a perforated vertical pipe.

For further particulars see **Mitchell's** "Mineral and Aerated Waters."

3. **By the Action of Acids on Carbonates.**—Carbon dioxide is usually prepared in the laboratory by the action of dilute acids on various carbonates:—

$$CaCO_3 + 2HCl = CaCl_2 + H_2O + CO_2.$$
$$MgCO_3 + H_2SO_4 = MgSO_4 + H_2O + CO_2.$$
$$2NaHCO_3 + H_2SO_4 = Na_2SO_4 + 2H_2O + 2CO_2,$$

and the same method is used industrially, though to a very much less extent now than formerly.

Both sulphuric acid (oil of vitriol) and hydrochloric acid are used, but the former more especially, because the volatile impurities contained in the latter acid render the purification of the resulting gas more difficult. The sulphuric acid must, however, be free from arsenic (a frequent impurity) in the case of carbon dioxide for mineral waters, etc.

As concerns the carbonates, whiting (purified chalk, $CaCO_3$) or sodium bicarbonate ($NaHCO_3$) has been usually employed in England, whilst powdered marble ($CaCO_3$) has been used in America. In Germany the mineral water manufacturers have generally employed magnesite ($MgCO_3$). Sodium bicarbonate with sulphuric acid yields very pure carbon dioxide, and has the advantage of leaving a soluble salt in the generator, but its use is very expensive. Marble frequently contains bituminous matter and iron, so that the evolved gas needs careful purification. Whiting, on the other hand, requires to be mixed with more water than powdered marble, so that a larger generator is required for the same amount of carbon dioxide; the action is somewhat violent, and the gas contains impurities. Both marble and whiting, with sulphuric acid, leave in the generator a troublesome mass of insoluble calcium sulphate or gypsum ($CaSO_4$), thus—

$$CaCO_3 + H_2SO_4 = CaSO_4 + H_2O + CO_2.$$

Magnesite has the advantage of giving a high yield of carbon dioxide (52 per cent. by weight), and the mother liquors can be worked for magnesium sulphate or Epsom salts ($MgSO_4$).

Concentrated acid to an amount generally less than that sufficient completely to neutralise the carbonate is run into a generator containing the carbonate mixed with water. The generators are usually constructed of lead, or cast iron lined with lead. Their form is that of a cylinder, either upright or, preferably, horizontal. The acid is admitted from an acid tank by means of a syphon tube, which acts also as a safety valve. An additional safety valve may be provided, especially in the case of cast-iron generators for use with sodium bicarbonate. A stirrer is also provided, which rotates within the vessel and keeps its contents well mixed. The driving rod of the stirrer passes through a stuffing box, which renders it gas-tight, thus preventing the escape of carbon dioxide.

Carbon dioxide generated by these methods needs a certain amount of purification, which is effected by means of water-scrubbers. If, after this treatment, it still possesses an unpleasant odour, it is further purified by leading through a solution of potassium permanganate, which destroys organic compounds and sulphuretted hydrogen, and one of sodium bicarbonate, which retains acid fumes. The gas may finally be filtered through wood charcoal.

Sometimes manganese dioxide or potassium permanganate is directly added to the generator, when sulphuric, but not hydrochloric, acid is used.

4. **As a By-product in the Burning of Limestone.**—Large quantities of carbon dioxide are evolved in the burning of limestone and marble for lime—

$CaCO_3 = CaO + CO_2$[1]—and several patents have been taken out for collecting and purifying the gas from this source.

The evolved gas, which contains about 30 per cent. of carbon dioxide, the remainder being mainly carbon monoxide and nitrogen, is drawn off from the upper portion of the furnace, and water-scrubbed. It is then sufficiently pure for the purpose of sugar refining. If pure carbon dioxide is required the gas may be absorbed by potash-lye (potassium carbonate solution), and re-evolved as described below.

5. **Recovery of Carbon Dioxide from Furnace Gases.**—Furnace gases resulting from the combustion of coke or charcoal contain about 20 per cent. by volume of carbon dioxide, and several processes have been patented for recovering this product from them.

The usual method (**Stead's** process) is to absorb the gas by means of potash-lye (potassium carbonate solution), and afterwards to liberate it by heating the carbonated lye under reduced pressure—

$$K_2CO_3 + H_2O + CO_2 \rightleftarrows 2KHCO_3.$$

The plant consists essentially of a lye boiler, directly fired with coke. The combustion gases, after washing and cooling, are forced through towers containing hard coke, in which they are absorbed by cold lye passed down the towers, reaction taking place in the direction of the upper arrow in the equation given above. The carbonated lye is then pumped into the boiler and boiled, when the reverse reaction takes place, and the carbon dioxide is expelled. The lye is then used over again. Heat interchanges are arranged in the paths of the lye, so that part of the heat of the hot decarbonated lye from the boiler may be given up to the cold carbonated lye on its way to the boiler. The evolved carbon dioxide is dried by means of calcium chloride, and either passed into a suitable gasholder, or immediately liquefied.

Carbon dioxide of a high degree of purity can be obtained by this method, comparing favourably with that produced from sodium bicarbonate. The gas is much purer than that obtained from acid and whiting, and as the process is considerably cheaper, it has to a very large extent supplanted the former method.

Knowles (English Patent, 1,744, 1907) provides means for the dilution of the carbonated lye, which he finds facilitates its decomposition. For other improvements see English Patent, 3,106, 1907, and French Patent, 387,874, 1908.

6. **Recovery from the Exhaust Gas of Gas Engines (Suerth System).**—In North Germany, carbon dioxide is obtained in a somewhat similar manner to the foregoing from the exhaust gas of gas engines. Producer-gas, after being scrubbed, is exploded in the cylinders of the engine. The issuing gases, which are very hot, are used to heat the carbonated lye in a special boiler. The gas is then passed through scrubbers containing limestone over which water trickles. Sulphur dioxide and dust are thereby removed. Next the gas is pumped, under a pressure of three to five atmospheres, through a vessel containing potash-lye, provided with a number of baffle plates, under each of which the gas accumulates in turn, and is thus exposed for a long time to the action of the lye. The carbon dioxide is thus absorbed ($K_2CO_3 + CO_2 + H_2O = 2KHCO_3$), the process being completed in an absorption tower. The carbonated lye is then pumped into the boiler, the carbon dioxide is expelled by boiling ($2KHCO_3 = K_2CO_3 + CO_2 + H_2O$), and the purified gas is dried and liquefied.

LIQUID CARBON DIOXIDE

Carbon dioxide has a critical temperature of (approximately) 31.0° C., the corresponding pressure being 73 atmospheres. It follows, therefore, that the gas

[1] A further quantity of carbon dioxide is, of course, obtained from the coke used for burning the limestone.

can be liquefied at ordinary temperatures by pressure alone. About 50-60 atmospheres are required, whereas at 0° C. 36 atmospheres are sufficient. At 760 mm. pressure the boiling point of the liquid is -78.2° C.; this is shown in the following table, as well as the vapour pressures corresponding to other temperatures:—

Temperature, Degrees C.	-78.2	-25	-15	-5	+5	+15	+25	+35	+45
Pressure in atmospheres (1 atmos. = 760 mm.)	1.0	17.1	23.1	30.8	40.5	52.2	66.1	82.2	100.4

Faraday was the first to succeed in bringing about the liquefaction of carbon dioxide, which he did by allowing sulphuric acid to act on ammonium carbonate in a sealed tube bent so as to form an inverted "V," the gas being liquefied by its own pressure. **Thilorier**, in 1839, first prepared liquid carbon dioxide on a large scale. He used two cast-iron cylinders, in one of which the carbon dioxide was formed by the action of sulphuric acid on sodium bicarbonate. The gas was liquefied by means of its own pressure, and then distilled into the second cylinder. The apparatus was dangerous, several fatal explosions occurring. A similar apparatus, in which lead cylinders surrounded by copper jackets (**Mareska** and **Donny**) were employed, was found to be more reliable.

Liquid carbon dioxide is a colourless mobile fluid, having a specific gravity of 1.06 at -34° C., 0.947 at 0° C., and 0.726 at +22.2° C. It will be seen from these figures that the coefficient of expansion of the liquid is greater than that of any gas. It is almost insoluble in water, but readily dissolves in alcohol or ether.

Owing to its convenience and the speed with which it can now be manufactured, carbon dioxide is nowadays usually stored and sold in the liquid form. Since the temperature rises when the gas is subjected to pressure, most modern compressors are of the multi-stage type, in which the gas is submitted successively to two, three, or even four different and increasing pressures, being cooled by means of cold water between each stage.

The most efficient type of cooling apparatus is constructed of double pipes. The water flows through the inner tubes, and the gas passes along the annular space between the pipes in the opposite direction. A submerged condenser is usually employed for the final stage.

The cylinders in which the liquid is stored are not completely filled, a space being left for the liquid to expand, otherwise there is a danger of explosion. The presence of air also greatly increases the risk of explosion, only 1 per cent. of air left in the cylinder increasing the pressure in it by over 3 per cent., in the most favourable circumstances (see **Stewart**, *loc. cit.*).

SOLID CARBON DIOXIDE

When liquid carbon dioxide is allowed suddenly to evaporate, part of it solidifies owing to the decrease in temperature produced by the evaporation.

The solid can be most readily prepared by tying a bag of coarse canvas around the outlet tube of a cylinder of liquid carbon dioxide, which is inverted so that the valve is underneath. On opening the valve, the liquid very rapidly blows into the bag, part vaporises and escapes through the pores of the canvas, whilst the remainder solidifies owing to the decrease in the temperature, and collects inside the bag in the form of a white bulky crystalline powder resembling snow (**Landolt**, *Berichte*, 1884, Vol. XVII., p. 309).

On compression in a wooden mould a compact mass is obtained, which, owing to the very large heat of vaporisation required, does not readily evaporate in the air. It may be retained in a cardboard or wooden box, or preferably wrapped round with indiarubber tissue, then with cotton-wool and paper. The specific gravity of the solid is about 1.2. Its temperature is -79° C., but it may be gently touched by the hand without sensible cold, owing to the protective atmosphere of gaseous carbon dioxide around it. If, however, it is pressed to the skin it forms a blister. On mixing with ether or chloroform the temperature falls to about -90° C., the mixture forming an excellent one for refrigerating purposes. On reducing the pressure the temperature falls still further to about −110° C. Liquid carbon dioxide immersed in a tube in this mixture solidifies to an ice-like solid (**Mitchell** and **Faraday**), having a specific gravity, at −79° C., of 1.56.

TECHNICAL USES OF CARBON DIOXIDE

The uses of carbon dioxide in commerce are many and increscent. The more important are noted below.

Carbon dioxide is used extensively in the manufacture of sugar for removing lime from the limed sugar-cane juice. It finds important application in the manufacture of various carbonates, especially white lead (France and Germany) and sodium bicarbonate, and in the manufacture of alkali both by the **Solvay** or ammonia soda process, and the cryolite process, in the latter of which carbon dioxide is used to decompose the sodium aluminate produced.

Carbon dioxide is also used for carbonating and raising beer, and in the manufacture of aerated waters. Liquid carbon dioxide, purchased in cylinders, has largely displaced the use of gas prepared at the works from whiting and sulphuric acid in the latter, as well as in other, industries. As **Mitchell** points out (*loc. cit.*), this substitution has eliminated the employment in the factory of sulphuric acid with its attendant dangers. It has considerably reduced the labour bill, owing to the ease with which the cylinders may be handled, and has cut down the cost of materials by about 10 per cent. Moreover, the liquid carbon dioxide on the market is purer than that usually obtained from whiting and sulphuric acid.

Carbon dioxide has been found useful for protecting wines from moulds, etc. A method of preparing carbon dioxide with antiseptic properties, in which it is sterilised by means of ozone, has been described by **Bourchaud-Praceig** (French Patent, 388,970, 1907).

Liquid carbon dioxide is also now extensively used as a refrigerating agent, especially in the manufacture of mineral waters, which need to be thoroughly cooled for saturation with gas during the bottling process. Liquid carbon dioxide has several advantages over both ammonia and sulphur dioxide as a refrigerant. It cannot impart, like ammonia or sulphur dioxide, any objectionable flavour to mineral waters with which it may come in contact, but can be used to carbonate them after it has done its work as a refrigerant. It does not attack copper, and its cost is very much less than that of liquid ammonia. Consequently carbon dioxide refrigerating machinery is rapidly displacing ammonia plant and sulphur dioxide plant.

Still colder temperatures can be produced by means of solid carbon dioxide prepared from the liquid as described above.

By allowing liquid carbon dioxide to come in contact with molten steel in strong gas-tight chambers, enormous pressures are produced, which free the casting from bubbles and increase the density of the metal. Liquid carbon dioxide is, therefore, used for hardening steel, especially by the firm of **Krupp** in Germany.

Liquid carbon dioxide has further applications in the removal of boiler-scale (see German Patent, 55,241, 1890), and as a fire extinguisher.

Liquid carbon dioxide may be used for raising sunken ships, etc. **Dr W. Raydt** was the first to carry out experiments in this direction, and in 1879 succeeded in raising an anchor weighing 316 cwt. (German) from a depth of 40 ft. in the sea at Kiel, by means of a carbon dioxide balloon.

Carbon dioxide possesses therapeutic properties, baths of water supersaturated with the gas having been found to be mildly stimulating, and thus useful in treating nervous diseases. Such baths are most readily prepared by means of liquid carbon dioxide.

CHAPTER VII

MANUFACTURE OF NITROUS OXIDE

(Laughing Gas, Nitrogen Monoxide), N_2O

By Geoffrey Martin.

The usual process is to heat ammonium nitrate, NH_4NO_3, in retorts, when it decomposes, thus:—

$$NH_4NO_3 = N_2O + 2H_2O.$$

The substance begins to decompose at 170° C., and the heat must be carefully regulated (best by gas firing), otherwise explosions can occur.

It is important to use pure ammonium nitrate. If the temperature is too high, N, NH_3 and the very poisonous NO are produced. The gas must be purified by passing through solutions of ferrous sulphate, $FeSO_4$, caustic potash, KOH, and milk of lime. NO is caught by the $FeSO_4$; and is held back by the KOH and lime, which also retains any CO_2.

1 kilo NH_4NO_3 gives 182 litres N_2O.

See Baskerville and Stephenson (*Journ. Ind. Eng. Chem.*, 1911, **3**, 579) for a full account of its preparation and the requisite purity for use as an anæsthetic.

Lidoff (*Journ. Russ. Phys. Chem. Soc.*, 1903, **35**, 59) mixes the ammonium nitrate with sand, and washes the gas with ferrous sulphate solution, drying it with an emulsion of ferrous sulphate in concentrated sulphuric acid.

Smith and Elmore (D.R.P., 71,279 of 1892) heat dry KNO_3 with dry $(NH_4)_2SO_4$. The evolution of gas begins at 230° C. and ends at 300° C. Thilo (*Chem. Zeit.*, 1894, **18**, 532) uses $NaNO_3$ and heats to 240° C. Campari (*Chem. Cent.*, 1888, 1569) heats 5 parts $SnCl_2$, 10 parts HCl (sp. gr. 1.21), and 0.9 parts HNO_3 (sp. gr. 1.38), when gas is evolved. Pictet (French Patent, 415,594 of 1910) and Södermann (French Patent, 411,785 of 1910) obtain it from an electrically produced nitrogen-oxygen flame by rapid cooling.

Properties.—Colourless gas with pleasant odour and sweet taste. Density, 1.5301 (air = 1). 1 litre weighs 1.9774 g. at 0° C. and 760 mm. Coefficiency of expansion, 0.0037067.

The liquefied gas has density 1.2257 ($H_2O = 1$), and refractive index 1.193 at 16° C. Critical temperature, 35.4° C.; critical pressure, 75 atmospheres. Liquid boils at −88° C., and thereby partially solidifies (at −115° C.). Mixed with CS_2 evaporated *in vacuo*, a temperature of −140° C. is attained.

Burning oxidisable bodies (such as P, S, etc.) continue to burn in the gas as in pure oxygen. With H gas it forms an explosive mixture. Heat of formation, 21,700 calories.

Solubility in water:—

1 volume water	5° C.	10° C.	15° C.	20° C.	25° C.
dissolves	1.048	0.8778	0.7377	0.6294	0.5443 vols.

When breathed, nitrous oxide is a valuable anæsthetic for short operations; 22-26 litres of gas are needed to produce insensibility. Prolonged breathing causes death. It is advisable to mix 0.1 per cent. of atmospheric air with the gas; the limit is 0.25 per cent. air. Mixed with oxygen the breathing of the gas produces intoxication. Also used for making **sodium azide** (see **Martin's** "Industrial Chemistry," Vol. II.).

Analysis.—Best by burning with H according to Bunsen's method. See W. Hempel (*B.*, **15**, 903; *Journ. Soc. Chem. Ind.*, 1882, 200). A. Wagner (*Journ. Soc. Chem. Ind.*, 1882, 332) gives a method of estimating the NO in N_2O. See also Lunge, *Journ. Soc. Chem. Ind.*, 1881, 428.

Transport.—In liquefied form in iron, steel, or copper cylinders. See R. Hasenclever, *Chem. Ind.*, 1893, 373.

CHAPTER VIII

The Ammonia and Ammonium Salts Industry

CHAPTER VIII

THE AMMONIA AND AMMONIUM SALTS INDUSTRY

BY GEOFFREY MARTIN, PH.D., D.SC.

For Synthetic Ammonia see p. 94.

Ammonia and Ammonium Salts.—Ammonia, NH_3, is a decomposition product of organic matter, resulting either from its destructive distillation or from its putrefaction and decay.

Until quite recently, practically the total supply of ammonia was obtained as a by-product in the destructive distillation of coal for the purpose of making illuminating gas, much also being produced by coke ovens, and the processes now worked for the gasification of coal, peat, oil-shale, etc.

The amount of nitrogen in coal amounts to 1-2 per cent., and only a relatively small proportion of this (from 12-20 per cent. of the total nitrogen present) is converted into ammonia during the process of destructive distillation or gasification. At least half of the nitrogen remains in the residual coke, and is not expelled completely even at a white heat. Much nitrogen escapes from the coal in the form of N_2 gas, being formed by the decomposition of the NH_3, which begins to decompose at 600°-800° C. A small proportion of the nitrogen distils over in the form of cyanide, pyridine, and other organic nitrogenous substances. The following figures show how the nitrogen contained in various sorts of coal is disposed of during the ordinary process of destructive distillation of 100 parts of their total N. We have :—

Kind of Coal.	English.	Westphalian.	Saar.
	Per Cent.	Per Cent.	Per Cent.
N in coke - - -	48-65	30-36	64
N as NH_3 - - -	11-17	11-15	16
N as gaseous N_2 - -	21-35	47-55	16
N as cyanide - - -	0.2-1.5	1-2	} 4.0
N in tar - - -	1-2	1-1.5	

In general, 100 kilos of average coal on destructive distillation yield 0.25-0.3 kg. of ammonia or 1.0-1.2 kg. of ammonium sulphate. When, however, the coal undergoes destructive distillation *in a stream of superheated steam* (as in the processes of producing "water-gas" or "Mond-gas" from crude coal) we can obtain as much as 3 kilos of ammonium sulphate per 100 kilos of coal treated.

The following may be considered as the chief sources of ammonia and ammonium salts, as they at present exist:—

(1) **The Coking of Coal.**

(*a*) **For the Production of Metallurgical Coke.**—Already in Germany no less than 90 per cent. of total output of ammonium sulphate is manufactured in recovery ovens; in England some 60 per cent., and in the United States only 20 per cent. is thus obtained. About 55 per cent. of the world's total output of ammonia in 1911, and 74 per cent. of that of the U.S.A., is derived from the carbonisation of coal in by-product ovens. There is little doubt that the output will increase as the amount of coal carbonised increases.

(*b*) **For the Production of Coal-Gas.**—30 per cent. of the world's output in 1910 was obtained from the retorting of coal for the manufacture of illuminating gas.

(2) **The Distillation of Shale.**—Large amounts are recovered in Scotland by the distillation of bituminous shale. The following figures, taken from the "49th Annual Report on Alkali, etc., Works," p. 130, show this:—

Year.	Total Shale Mined and Quarried in Scotland.	Total Sulphate of Ammonia Recovered from the Shale in Paraffin Oil Works.	Yield in Lbs. per Ton of Shale.
	Tons.	Tons.	
1903	2,009,265	37,353	41.6
1904	2,331,885	42,486	40.8
1905	2,493,081	46,344	41.6
1906	2,545,724	48,534	42.7
1907	2,690,028	51,338	42.7
1908	2,892,039	53,628	41.5
1909	2,967,017	57,048	43.1
1910	3,130,280	59,113	42.3
1911	3,116,803	60,765	43.7
1912	3,184,826	62,207	43.7

Considerable undeveloped deposits of bituminous shale exist in Newfoundland, Australia, and other parts of the world.

(3) **The Distillation or other Treatment of Peat.**—Much nitrogen is combined in peat (which, therefore, has found some application as a fertiliser), and many processes are either worked or are projected for directly transforming this nitrogen into the form of ammonium sulphate. The destructive distillation of peat has been proposed by Ziegler and others.

More promising is the partial combustion of peat with production of producer-gas (Frank, Caro, Mond, described on p. 43) or the slow wet-combustion of peat (Woltereck, p. 43). There are 20,000 million tons of undeveloped peat in the U.S.A., while equally enormous quantities exist in Ireland, Canada, Newfoundland, Sweden, Norway, Russia, Germany (Prussia alone containing 5,000,000 acres).

The average nitrogen content of these peat deposits may be taken as 2.05 per cent. (sometimes reaching 4 per cent. in the case of dry peat), so that if only 50 per cent. of this was recovered, very large amounts of ammonium sulphate would become available.

(4) **Producer-Gas.**—The ammonia contained in the coals used in producing this gas—the main type of plant used being the Mond Gas-Producer—is now very large and is likely to increase. The process is described in **Martin's** "Industrial Chemistry," Vol. II.

(5) **Blast-Furnace Gas.**—The nitrogen in the coal used in blast-furnaces escapes in part as ammonia, much, however, being decomposed by the high temperatures in the furnace.

Some of the combined nitrogen, however, comes from the nitrogen of the air by actions taking place inside the furnace, principally the formation of cyanides, which are then decomposed into ammonia by aqueous vapour in the furnace. In the United Kingdom, for 1910, about 20,130 tons of ammonium sulphate were recovered, 20,130 tons being recovered in 1911, and 17,026 in 1912.

(6) **Production from Beetroot Sugar Waste, "Schlempe" or "Vinasse."**—The thick brown liquid remaining after the extraction of all the possible sugar from the syrup is known as "vinasses" or "schlempe."

It contains much nitrogen and potassium salts. Until recently it was the custom merely to calcine this material so as to obtain the potash salts in the form of "schlempe kohle." Bueb of Dessau and Vennator now recover nitrogen from this by distilling the schlempe from iron retorts, leading the evolved vapours through chambers filled with brick cheque-work maintained at a red heat, whereby the complex vapours decompose into HCN, NH_3, etc.

The ammonia and cyanogen are then recovered as in the purification of coal-gas. The process is described in **Martin's** "Industrial Chemistry," Vol. II.

(7) **By Distillation of Bones, Leather, and other Nitrogenous Organic Matter.**—The distillation of bones, for the production of "bone black," formerly yielded a considerable supply of ammonia, but the industry is now not so prominent as formerly.

The ammonia may be extracted from the evolved gases by scrubbing, as in coal-gas manufacture. The treatment of animal refuse for the manufacture of prussiate is now obsolete. It is more profitably employed as manure. See **Martin's** "Industrial Chemistry," Vol. II.

(8) **From Sewage and Urine.**—A very rich source of ammonia is ordinary urine. 100,000 heads of population could produce per year about 6,000 tons of NH_3. If all the ammonia corresponding to London urine were collected, more than 60,000 tons of ammonium sulphate could be annually produced therefrom.

The method of collection of urine and its working up into ammoniacal compounds has been carried on at Paris and at Nancy. In 1909 France obtained 13,000 tons of ammonium sulphate therefrom, 10,000 being obtained in Paris alone. However, the collection and utilisation of animal excrement is so nauseous and costly and dangerous a process, that the bulk of the enormous ammonium supplies producible from this source are run to waste.

The process consists in allowing the urine to ferment into ammonium carbonate. The clear liquor is distilled and the ammonia recovered as in gas-liquor.

For further details see Ketjen, *Zeit. angew. Chem.*, 1891, 294; Butterfield and Watson, English Patent, 19,502, 1905; Taylor and Walker, U.S. Patent, 603,668; Young, English Patent, 3,562, 1882; Duncan, German Patents, 27,148, 28,436.

(9) **Synthetic Ammonia.**—Enormous supplies of ammonia are now becoming available by the synthesis of ammonia, either directly from atmospheric N and H, or else from cyanamide or nitrides. These processes are discussed in detail in a separate article, p. 94.

The production of ammonium sulphate is increasing rapidly in order to meet the increasing demand for nitrogenous manures. The following figures refer to Great Britain:—

AMOUNT OF AMMONIA RECOVERED IN THE UNITED KINGDOM (EXPRESSED IN TERMS OF SULPHATE)—TONS.

	1910.	1911.	1912.	1913.	1914.
Gas works	167,820	168,783	172,094	182,000	177,000
Iron works	20,139	20,121	17,026	20,000	19,000
Shale works	59,113	60,765	62,207	63,000	62,000
Coke-oven works	92,665	105,343	104,932	...	...
Producer-gas and carbonising works (bone and coal)	27,850	29,964	32,049	167,000	163,000
Total	367,587	384,976	388,308	432,000	421,000

The following figures refer to the output of the chief countries:—

	1900.	1909.	1911.
	(Metric) Tons.	(Metric) Tons.	(Metric) Tons.
England	217,000	349,000	378,000
Germany	104,000	323,000	400,000
United States	58,000	98,000	127,000
France	37,000	54,000	60,000
Belgium, Holland, Austria, Russia, etc.	68,000	134,000	...

The **world's production** of ammonium sulphate is estimated as:—

1900.	1909.	1911.
484,000 tons.	950,000 tons.	1,150,000 tons.

Great Britain exported ammonium sulphate:—

1911.	1912.	1914.
292,000 tons.	287,000 tons.	314,000 tons.

The chief product of the ammonia industry is, at present, *solid ammonium sulphate*, $(NH_4)_2SO_4$, which is principally used in agriculture as a manure. It is valued on the percentage of nitrogen it contains, containing when pure, as $(NH_4)_2SO_4$, about 21.2 per cent. N against 16.5 per cent. N in $NaNO_3$, or Chile saltpetre, which is at present its great competitor as a manure.

The price of ammonium sulphate sunk from 50s. per ton in 1880 to 20s. to 30s. in 1909, and no doubt in consequence of the production of cheap synthetic ammonia, both from cyanamide and by direct synthesis, the price will probably still further decrease.

Manufacture of Ammonium Sulphate from Gas-Water or Ammoniacal Liquor.—At present the bulk of the ammonium sulphate on the market is derived from the "gas-water" or "*ammoniacal liquor*" produced in the numerous coal-gas producing plants, coke ovens, etc.

In the coke ovens, however, the gases now are passed directly through sulphuric acid, and the ammonium sulphate thereby directly fixed, thus avoiding the initial production of an "*ammoniacal liquor*" such as is indispensable to coal-gas production.

Ordinary ammoniacal liquors contain some 1.5-3 per cent. NH_3, united with various acids. The chief acid is carbonic, H_2CO_3, but besides this we get H_2S, HCN, $HCNSH_2S_2O_3$, H_2SO_4, HCl, and ferro- and ferri-cyanic acids.

The ammonium salts are, in practice, divided into (*a*) volatile, (*b*) fixed. The "volatile" ammonium salts on boiling with water dissociate, evolving ammonia. The chief volatile salts are:—Ammonium carbonate, $(NH_4)_2CO_3$; ammonium sulphide, $(NH_4)_2S$ and NH_4HS; ammonium cyanide, NH_4CN.

The "fixed" ammonium salts (*e.g.*, ammonium sulphate, $(NH_4)_2SO_4$; ammonium chloride, NH_4Cl, etc.), are not decomposed by water, but the ammonia has to be driven out of them by boiling with milk of lime.

Different ammoniacal liquors, however, have an extremely variable composition. An average sample would contain per 100 c.c. from 1.4-3.3 g. volatile ammonia (principally in the form of ammonium carbonate), and 0.2-0.6 g. NH_3 in the form of "fixed" salts. The composition of the ammoniacal liquor naturally largely depends upon the nature of the coal used, and some coals, rich in chlorides, yield ammoniacal liquors containing much NH_4Cl.

The following analyses give the composition of some average ammoniacal liquors, the numbers giving grams per 100 c.c.:—

	Gas Works.	Coke Ovens.	Blast Furnaces.	Shale Works.	Coalite Works.
Volatile NH_3	1.4-3.3	0.84	0.2-0.4	0.9	1.5
Fixed NH_3	0.6-0.2	0.10	0.008-0.009	0.03	0.17
Total NH_3	2.5-3.5	0.94	0.2-0.4	0.9	0.7
$(NH_4)_2S$	0.9-0.8	0.47	...	0.1	0.23
$(NH_4)_2CO_3$	5.0-8.8	1.96	1.1	2.9	6.4
$(NH_4)Cl$	1.1-0.5	0.22	0.006	0.015	0.1
$(NH_4)_2SO_4$	0.2-0.0	0.03	0.009	0.016	0.05
$(NH_4)_2S_2O_3$	0.17-0.0	0.04	0.002	0.09	0.4
NH_4CNS	0.53-0.07	0.04	0.003	...	0.3
NH_4CN	0.036-0.07	0.07	0.003	...	...
$(NH_4)_4Fe(CN)_6$	0.038	...	...	...	...

Organic substances such as phenol, pyridin, acetonitrol, etc., also occur in small amounts.

The method of working the ammoniacal liquor for ammonium sulphate is first to boil it until all the "volatile" ammonium salts have been distilled off. To the residual liquid containing the "non-volatile" ammonium salts the theoretical amount of milk of lime is added, and the boiling continued until all their ammonia is also expelled.

The evolved vapours are usually led directly into H_2SO_4 of 42°-46° Bé. (81°-93° Tw.), and the ammonia fixed in the form of solid ammonium sulphate, $(NH_4)_2SO_4$, which can be sold without further refining for manurial purposes.

The plant used consists of "column" apparatus similar to those described in detail in **Martin's** "Industrial Chemistry," Vol. I., for distilling alcohol; the apparatus, however, is modified so that the ammoniacal liquor alone is distilled in the upper part of the apparatus so as to expel all volatile ammonia; while in the lower part of the apparatus the "fixed" ammonium salts in the residual liquid are boiled with milk of lime.

There are a great many different plants on the market, some of which are extremely efficient. **Feldmann's Apparatus** (D.R.P., 21,708, see English Patent, 3,643, 1882) is shown in Fig. 38.

The ammoniacal "gas-water" flows into a tube from the regulating tank A

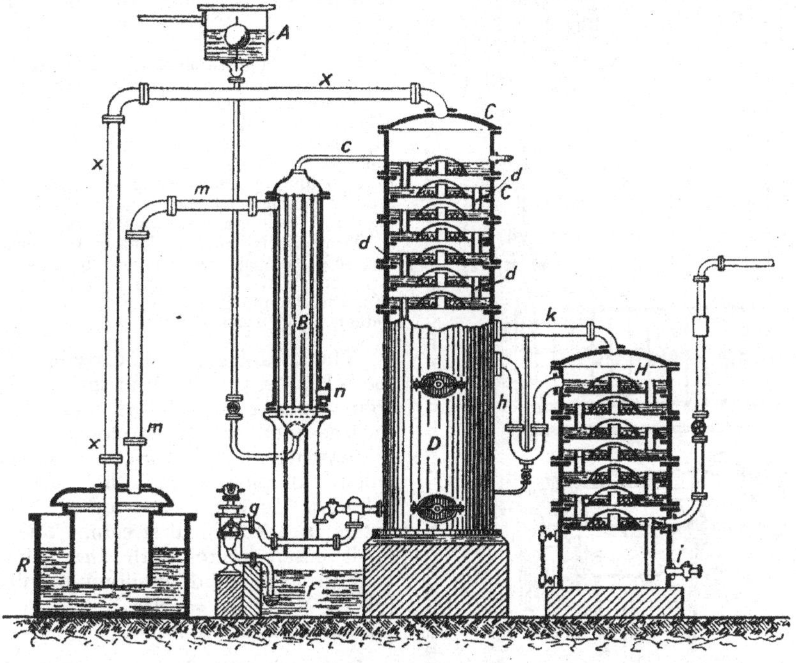

FIG. 38.—Feldmann's Ammonia Still.

and enters the multitubular "preheater" B, consisting of a series of tubes through which the ammoniacal liquor flows, which are themselves heated by the steam and hot gas coming from the saturator R by the pipe MM. From the "preheater" B the now hot ammoniacal fluid flows into the top chamber of the column C. This is provided with a number of compartments each provided with an overflow pipe D, so that in each compartment the liquor accumulates to an appreciable depth. In the centre of the floor of each compartment is a wider pipe covered over with a "bell" or "mushroom" (*e*), provided with serrated edges (see **Martin's** "Industrial Chemistry," Vol. II., under "Ammonium Soda Industry"). Through this central pipe the ammoniacal gases and steam come up from below and stream through the liquor surrounding the "mushroom," and thus boil out all the volatile NH_3.

The liquid in C, from which all volatile ammonia has been boiled out, now enters the lower part of the still D. Into this compartment a stream of milk of lime is continually pumped by means of the pump *g*, the lime being sucked out

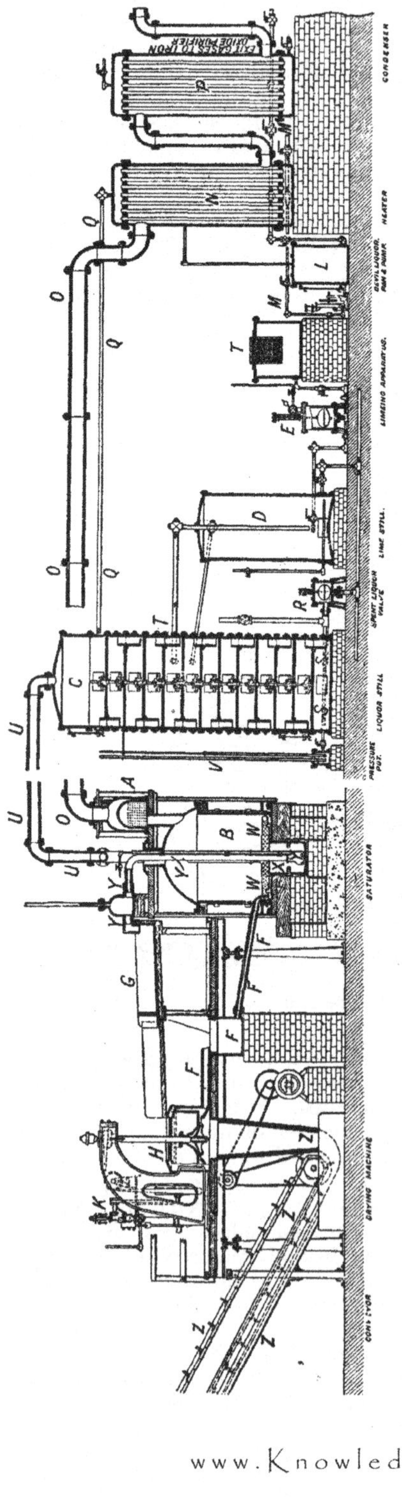

FIG. 39.—Ammonium Sulphate Plant of the Chemical Engineering Co., Hendon, England.

of the tank *f*. The mixed fluids flow through a filtering sieve (to retain large particles) through the tube *h* into the column H, through which is passed from below a current of steam from a boiler. This steam maintains the whole liquid at a boiling temperature and completely expels all the ammonia from the "fixed" ammonium salts, the ammonia being, in the first place, set free by the milk of lime.

The waste ammonia-free liquors run away through *i*.

The mud-like mass of lime escapes through an opening at the bottom of D.

In some plants a special separate mixing vessel is provided, standing outside the column. Into this the liquid coming from the middle part of the column C is run, and is then intimately mixed with milk of lime, and then the mixed fluids are run back into the lower column B, and subjected to the boiling by means of steam. A still of this type is manufactured by the Chemical Engineering Co., Hendon, and is shown in Fig. 39.

Many special modifications of this apparatus are used. We may here mention the apparatus of Grüneberg and Blum (D.R.P., 33,320); Wilton (English Patent, 24,832, 1901); Scott (English Patent, 3,987, 1900; 11,082, 1901).

The escaping steam, carrying with it the NH_3 gas, passes out at the top of the column C and through the tube X X into the lead-lined (or volvic stone) "*saturator*" R constructed as shown, with a leaden "bell" dipping under the surface of sulphuric acid (90° Tw. 45° Bé.) which enters in a continual stream. The NH_3 as it enters unites with the sulphuric acid to form solid ammonium sulphate ($2NH_3 + H_2SO_4 = (NH_4)_2SO_4$) which separates out in the liquid.

It should be noted also that much heat is evolved by the union of the sulphuric acid and the ammonia in the saturator, the heat of interaction being sufficient not only to maintain the saturator at the boiling point, and compensate for unavoidable losses by radiation, etc., but also to more than evaporate the whole of the water contained in the acid, amounting to 20-30 per cent., so that the apparatus can be washed out from time to time, as necessary, without wasting the washings, or evaporating them down externally.

The saturator just described belongs to the "*partly open*" type, the hot waste gases being led off by the pipe M to the heater B, and the sulphate accumulating on the floor of the saturator, the workman removing the latter as it separates by means of a perforated ladle inserted through the open part of the tank. The crystals of sulphate are then placed on a lead-lined drainer, so that the mother liquors flow back to the saturator.

More often the ammonium sulphate crystals are removed from the saturator by means of a steam discharger (working on the principle of the air lift), which drives it in the form of a coarse mud, together with a considerable amount of mother liquor, on to the drainer, and thence it is passed into the centrifugal machines, the mother liquors invariably running back into the saturator. In this type of plant a closed saturator is used, as described below:—

Fig. 39 shows a modern ammonium sulphate plant, erected by the Chemical Engineering Co., of Hendon, London, N.W., for the Grangetown Gas Works, Cardiff, in 1911. This plant is capable of producing 3 tons of ammonium sulphate per twenty-four hours.

The ammoniacal liquor enters the multitubular heater N at the bottom through H, passes up the pipes' insides, and is thereby heated to the boiling point by the steam and hot waste gases from the saturator B. These hot vapours issue from the saturator at A, pass along the pipe OO, then encircle the tubes in the interior of the heater N, heating the ammoniacal liquor therein contained to the boiling point, and being partially condensed in so doing. The condensed steam (known as **Devil Liquor**, on account of the H_{23}, NH_3, HCN, etc., contained therein) runs off into the sealed closed tank L, and is pumped back through MM, and mixed with the gas liquor, the whole thus passing through the liquor still C, which renders it inodorous, the liquors being finally run to waste after passing through the liquor still in the form of "spent liquor." Any moisture not condensed by the heater N is finally condensed in the condenser P (two are employed) of similar construction to N. The H_2S and CO_2 from the saturator B, however, pass for the most part away from the exit gases, being purified with oxide of iron, and the H_2S recovered is sulphur.

The ammoniacal liquor, heated to boiling in N, passes along the pipe QQ, and enters the liquor still C at the top, and slowly flows by a series of weirs, in a downward direction, through the fifteen chambers to the automatic exit, which is controlled by the spent liquor valve R. V is an arrangement for controlling the pressure in the still. Steam is admitted at the bottom of the still by the perforated pipe SS, and bubbles through the liquor in each chamber, travelling in a reverse direction to the liquor, and carrying with it the ammonia and the gases, H_2S and CO_2, which pass away from the top of the still to the acid saturator B; the H_2S and CO_2 thence escape along OO to the condensers, and reappear in the "devil liquors" at N, as previously explained, and in the waste exit gases from P.

In order to complete the removal of fixed ammonium, salts from the liquor lime are automatically admitted to the middle chamber of the still at T, D being the auxiliary limeing still (see p. 84), which acts as a reservoir, and retains the lime sludge. The lime is slaked in T with spent liquor from the stills, and forced into the still C at boiling temperature at the required rate, by means of automatic pumps. E is the automatic valve controlling the admission of lime to D. The saturator B is of the round closed type, constructed of 40 lbs. chemical lead. In this saturator the ammonia coming from the still C passes through the pipe UU nearly to the bottom of the saturator, and bubbles through the acid bath by means of the perforated pipe WW. The saturator is continuously fed with sulphuric acid in proportion to the amount of entering ammonia. The chemical action, caused by the ammonia uniting with the acid to form ammonium sulphate, develops sufficient heat to cause vigorous boiling; the uprising steam, together with H_2S and CO_2 from the ammoniacal liquors, and some ammonia escape at a high temperature through the tube OO to the heater N and condensers P, as above explained.

The sulphate of ammonia is deposited in a crystalline form into the well X at the bottom of the saturator B, from which it is pumped by means of the steam discharger YY to the receiving tray G, whence it gravitates at intervals to the centrifugal machine (described in **Martin's** "Industrial Chemistry," Vol. I.), the tray holding a charge of about 4 cwt. The centrifugal machine separates the mother liquor, which flows back to the saturator by FFF. After two minutes' spinning the centrifugal machine H is stopped, and the dry sulphate of ammonia is dropped through the centre valve on to the elevator conveyor ZZ, and deposited in the store.

The apparatus used for the manufacture of ammonium sulphate by the Coppée Gas Co. is shown in Fig. 40. The diagram, Fig. 41, explains the mode of working of the apparatus.

The crystals of sulphate thus obtained are sometimes washed with a very little water, dried, and put on the market for manurial purposes, containing 25.1-25.3 per cent. NH_3. $(NH_4)_2SO_4$ requires $NH_3 = 25.8$ per cent. less than 0.4 per cent. free H_2SO_4, and no cyanide, since the latter is very injurious to vegetation.

Chemically pure ammonium sulphate is sometimes, although rarely, obtained from this raw ammonium sulphate by crystallisation.

Treatment of the Waste Exit Gases from the Ammonium Sulphate Plant.—The acid gases, such as H_2S, HCN, and CO_2, which pass with the ammonia into the sulphuric acid in the

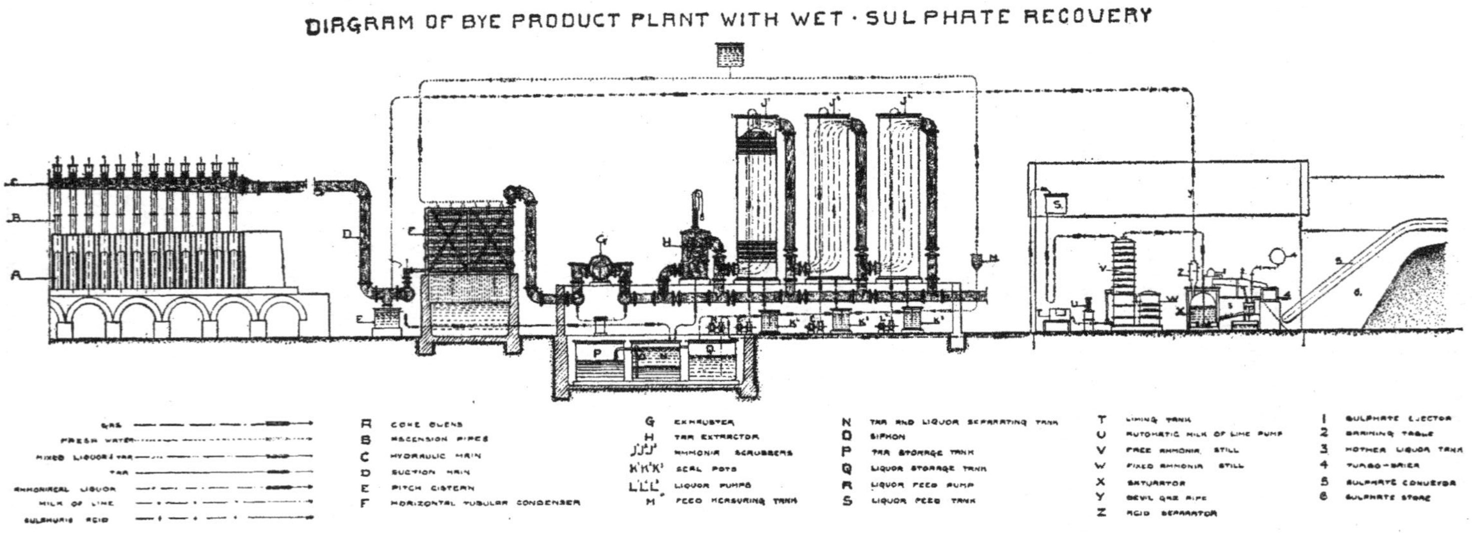

THE COPPÉE COMPANY (GREAT BRITAIN) LTD., KING'S HOUSE, KINGSWAY, W.C.

FIG. 40.—By-Product Plant with Wet Sulphate Recovery.

saturator are not absorbed therein, but escape with steam into the multitubular preheaters and condensers, and after heating the entering ammoniacal liquors in the tubes in the preheater as they pass on their way to the still, the cooled gases consisting of H_2S, HCN, and CO_2, finally escape either (1) directly into a furnace, where they are burnt in order to destroy the poisonous HCN, the H_2S burning to SO_2; or (2) into an **absorption apparatus** using iron oxide purifiers (as in coal-gas works for purifying coal-gas) (see **Martin's** "Industrial Chemistry," Vol. I.); or (3) where the H_2S is recovered as sulphur, and nitrogen recovered as Prussian blue, the gas is burnt in a limited supply of air according to the **Claus** process, whereby the H_2S burns, depositing S, which is thereby recovered.

However, in most works the gases are simply burnt, the products of combustion escaping, unutilised, up the chimney.

MANUFACTURE OF AMMONIUM SULPHATE FROM MOND GAS.[1]

Very similar is the production of ammonium sulphate from **Mond Gas.**

The coal is charged into a distributor and hopper from an elevator, and then falls into the "generators" where, under the influence of a mixture of air and steam, it is gasified in the manner described in Chapter V. See also the section on "Gaseous Fuels," **Martin's** "Industrial Chemistry," Vol. II., the equations being:—

$$C + O_2 = CO_2. \qquad CO_2 + C = 2CO. \qquad H_2O + C = CO + H_2.$$

There is thus produced a gas consisting mainly of CO, admixed with some hydrogen. The N in the coal escapes in 70-80 per cent. as NH_3 with the issuing gas. The gases next stream through a series of "coolers," in which they exchange their heat with the air which is streaming into the generators, the air being thus preheated, and the Mond gas cooled. Next, the gas passes through a washer, where it is largely freed from tar and dust. Finally, the gas passes up a tower, where it meets a stream of descending sulphuric acid, which combines with all the ammonia in the gas, forming ammonium sulphate. This acid solution is pumped by an acid pump up a tower, a continual circulation of the acid in the tower being kept up until the acid is practically saturated with ammonium sulphate, when it is run off and evaporated in a special apparatus, and the ammonium sulphate recovered. From the tower the gases pass through another tower, where they are treated with a stream of cold water, whereby the gas is cooled (and the water heated), and passes away directly to the gas engines or furnaces for use. The hot water obtained from the second tower is now pumped up the third tower, where it is allowed to flow down against an incoming current of cold air, which it thus saturates with water vapour, the air current then going on to the furnaces for the production of the CO by partial combustion of the coal.

While in coal-gas works and coke ovens scarcely ever more than 20 per cent. of the N in the coal is obtained in the form of ammonia, in the Mond-gas process no less than 70-80 per cent. of the nitrogen present in the coal is ultimately converted into ammonia, and recovered as ammonium sulphate, *e.g.*, each ton of coal yields over 40 kg. of ammonium sulphate, against 10 kg. obtained in coke ovens. Over 4s. per ton profit can be made out of the ammonium sulphate thus recovered, which leads to a further reduction in the price of the gas for power.

According to Caro, even the waste obtained by washing certain coals—containing only 30-40 per cent. C and 60-70 per cent. ash, and so useless for burning in the ordinary way—can be gasified by the Mond process, and a very considerable percentage of the nitrogen recovered as ammonium sulphate, 1 ton of this waste material yielding 25-30 kg. of ammonium sulphate and 50-100 H.P.-hours in the form of electrical energy.

Also **moist peat** can be gasified in the generators, and the contained nitrogen recovered as ammonium sulphate (Woltereck).

For further details the reader should see "Industrial Chemistry," Vol. II., "Gaseous Fuels"; also Caro, "Die Stickstoffrage in Deutschland" (1908); *Chemiker Zeitung*, 1911, 505; Frank, Caro, and Mond, *Zeit. angew. Chem.*, 1906, 1569; Norton, *Consular Report*, pp. 40, 170; Lange, "Coal Tar and Ammonia," 1913, 861; Woltereck, English Patents, 16,504, 1904; 28,963, 1906; 28,964, 1906.

MANUFACTURE OF AMMONIUM SULPHATE BY THE DIRECT PROCESS FROM COKE-OVEN GAS, BLAST-FURNACE GAS, PRODUCER-GAS, AND SIMILAR GASES RICH IN AMMONIA

This is a problem on which a great deal of elaborate work has been expended during the last fifteen to twenty years, and even at the present time it does not seem to have been completely solved.

Ammonium carbonate is very volatile, and consequently passing the gases directly through water only causes the formation of a dilute ammoniacal "gas liquor" (as in gasworks), the direct distillation of which would be expensive on account of fuel consumed.

It is, therefore, much more economical to pass the gases directly into fairly concentrated sulphuric acid, whereby matters must be so arranged that solid ammonium sulphate separates directly, that the concentrated sulphuric acid is not greatly diluted by the steam, etc., in the gases driven through it, and finally, that the resulting tar is not spoiled by the treatment.

One of the most successful systems is that embodied by the **Kopper Ammonia**

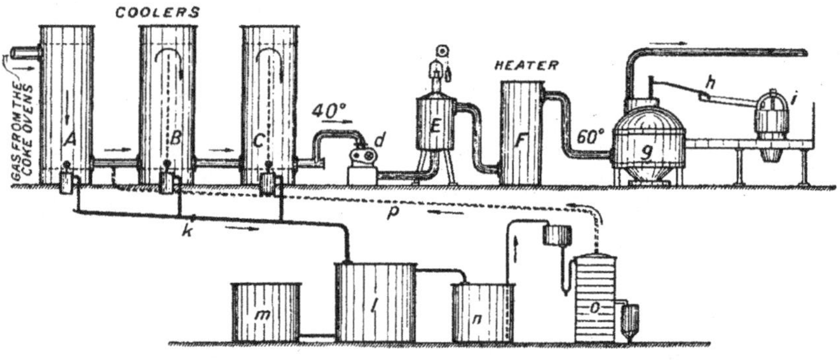

FIG. 42.—The Kopper Ammonia Recovery Plant.

Recovery Plant. The gases coming from the coke ovens, etc., are first cooled to 40° C., whereby the heavy tar oils and the bulk of the steam (with 20-25 per cent. of the total ammonia) is deposited in the liquid form.

The gases are next passed through tubes whereby they are reheated to 60°-80° C.—the hot furnace gases in counter-current being used for this purpose—and the hot gases are then directly passed into 60° Bé., 141° Tw., H_2SO_4; simultaneously the NH_3 which has been expelled from the condensed liquors by heating them with lime is also passed into the H_2SO_4. The NH_3 directly combines with the H_2SO_4 and ammonium sulphate separates in a solid form in the saturator and is withdrawn from time to time, centrifuged, and dried as previously described.

Fig. 42 shows **Kopper's Plant.**

The hot furnace or coke-oven gases are passed through coolers *a*, *b*, *c*, until they are cooled to about 30°-40° C., and then are led through a Pelouze tar separator (**Martin's** "Industrial Chemistry," Vol. I.) *e*, so that almost all the tar (but not the light oils) are condensed, together with most of the water, which contains 25-75 per cent. of the total ammonia (according to the temperature), all the fixed ammonium salts (NH_4Cl, etc.), etc., etc. The tar and the ammoniacal water flows into a holder *l*, the tar being run off at the bottom into another holder *m*, while the ammoniacal water runs into a second holder *n*.

The tar-free gases are again reheated in *f* to 60°-80° C., and then are passed directly into the holder *g*, containing concentrated 60° Bé. (141° Tw.) sulphuric acid. The solid ammonium sulphate immediately separates, and is forced out from time to time by means of compressed air into the collecting tray *h*, thence into the centrifugal machine *i*, where excess of acid is drained off.

The reheating of the gas, before passing into the sulphuric acid, aids the evaporation of the water from the sulphuric acid. The heat of the reaction going on in *g* is alone often sufficient to achieve this, especially if all the fixed ammonium salts have been previously removed. The gases escaping from *g* contain benzene, and are passed on through heavy oils to extract this substance, as described under **Coke Ovens** in **Martin's** "Industrial Chemistry," Vol. II.

The ammoniacal water in *n* is distilled in a column with lime in the usual manner for ammonia (see p. 83), and the gaseous ammonia evolved is often directly sent back into the gas stream (between the coolers *a* and *b*) to be fixed by the sulphuric acid in *g*. Since the amount of deposited gas-water here only amounts to about 20 per cent. of the washing water formerly needed to extract the ammonia in scrubbers, it is stated that the cost of the distillation of the ammonia and the quantity of the troublesome waste water is much less than by the ordinary process of the ammonia extraction.

In the **Otto-Hilgenstock Ammonia Recovery Process** (Fig. 43) the old condensing plant is entirely dispensed with, the tar being removed from the entering gases by a tar spray at A at a temperature above the dew-point of the liquors.

After depositing the tar in B the gases pass directly through an exhauster E into the saturator F, where the whole of the ammonia is caught by the sulphuric acid. The gases coming from the saturator are hot, and contain all their moisture in the form of steam; the gases are, therefore, passed forward to the oven flues, and the troublesome and offensive waste liquors are thus got rid of. C is the tar-spray pump, D the tar-spray feed pipe, G is the acid-spray catch box, H the mother liquor return pipe, J the tar store, K the tar-spray overflow pipe, L the condensing tank, M the pump delivering tar to railway trucks, N the pump delivering condensers to the saturator F.

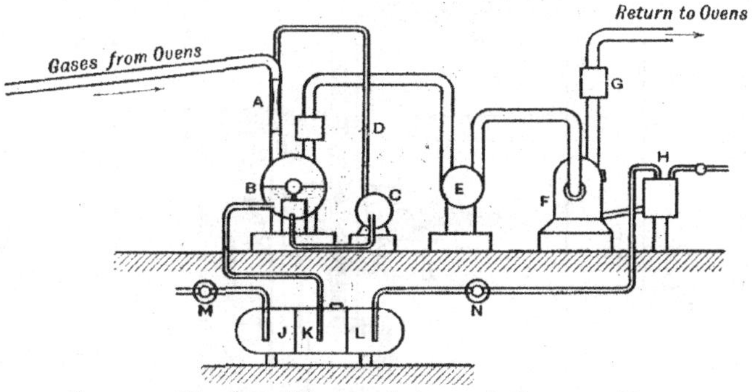

FIG. 43.—The Otto-Hilgenstock Ammonia Recovery Plant.

This process, by abolishing condensing plant, liquor tanks, ammonia stills, lime mixers, pumps, etc., effects a great saving, since less floor space is required, and nearly the whole of the steam required to distil the ammoniacal liquor made by the condensing process is abolished. Also no ammonia is lost, as often arises in a distilling plant. The ammonium sulphate produced contains 25-25.5 per cent. N and contains less than 0.1 per cent. of tar.

The Coppée Company's Process of Semi-Direct Sulphate Recovery is illustrated in Figs. 44 and 45.

MANUFACTURE OF CAUSTIC AMMONIA (LIQUOR AMMONIA)

A **crude** aqueous solution of ammonia, containing ammonium sulphide and sometimes carbonate, is manufactured under the name "**Concentrated Ammonia Water**" by distilling ordinary gas-water without addition of lime in column apparatus.

Concentrated Ammonia Water is used as a convenient source of ammonia either for ammonia-soda factories, or for factories which, possessing no ammonia plant of their own, require ammonia gas in a concentrated form, *e.g.*, for manufacturing salts such as NH_4Cl, NH_4NO_3, etc., or liquid NH_3. It contains much less CO_2 and H_2S than the ordinary diluted ammoniacal liquors of gas-works, because these gases, in general, escape during the distillation before the NH_3 gas. This liquor is much cheaper than the pure aqueous solution, and the presence of some CO_2 and H_2S does not affect the manufacture of many chemicals.

DIAGRAM OF BYE PRODUCT PLANT WITH SEMI DIRECT SULPHATE RECOVERY

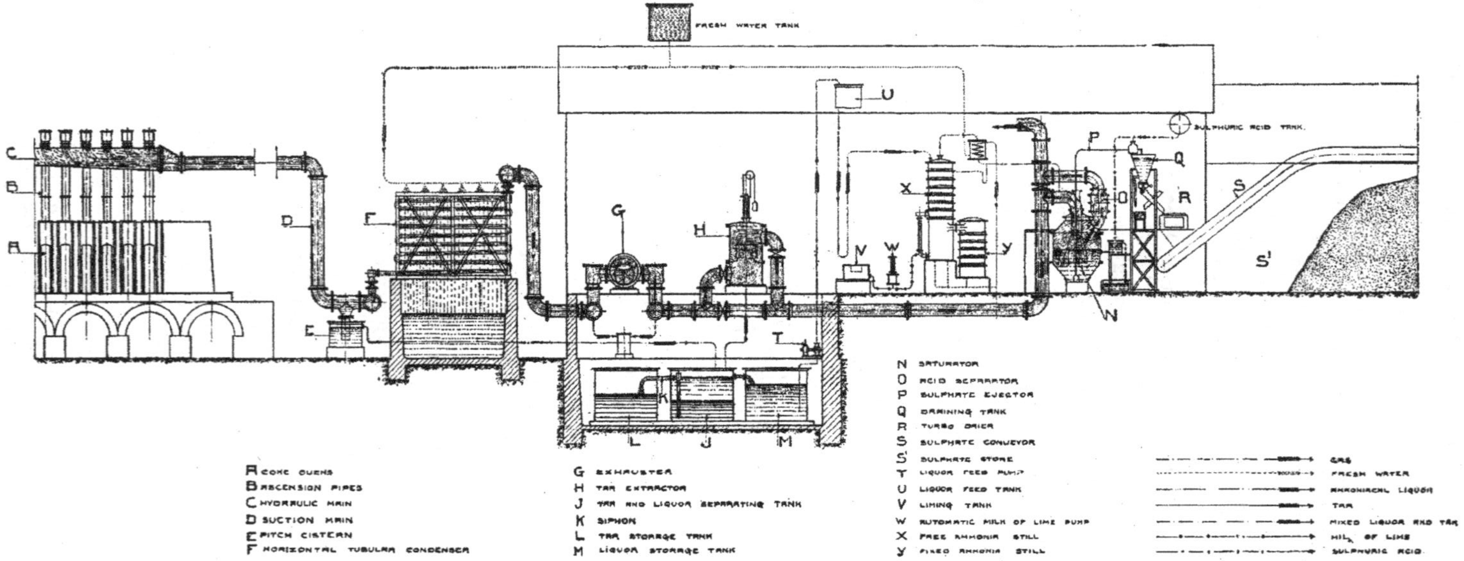

THE COPPÉE COMPANY (GREAT BRITAIN) LTD., KING'S HOUSE, KINGSWAY, W.C.

FIG. 44.—By-Product Plant with Semi-direct Sulphate Recovery.

Two varieties of "concentrated ammonia water" are manufactured:—(1) One containing 15-18 per cent. of NH_3 with both sulphide and carbonate present; (2) one containing 18-26 per cent. NH_3 containing some sulphide but practically no carbonate.

(1) The first liquid is made by passing the ammoniacal gases from the still (as described above) through a reflux condenser, whereby some moisture is removed, and then into a direct condenser, the gases from which are washed through water. The formation of ammonia carbonate and resulting blockage of the pipes prevents a higher concentration of NH_3 being obtained in this manner.

(2) To manufacture the second liquid we proceed as before, but the vapours from the still are passed first through the reflux condenser and then through vessels containing milk of lime which removes the CO_2 and some H_2S. The vapours are then condensed and contain 22-26 per cent. NH_3, some H_2S, but little or no CO_2.

Manufacture of Pure Aqueous Solutions of NH_3.—The ammoniacal liquor is first heated to 70°-80° C. in order to expel most of the CO_2 and H_2S. The liquid is next distilled in a columnar apparatus, as previously described, and the vapours are passed, first through a reflux condenser (to remove some water), then through milk of lime washers to remove CO_2 and H_2S (the partly used lime being run back into the still to decompose the fixed ammonium salts), the last traces of H_2S being removed by either passing the gases through $FeSO_4$ solution, or through a little NaOH solution. J. Louis Foucar recommends ammonium persulphate or sodium permanganate.

Lastly, the vapours pass through wood-charcoal, which removes tarry matters, and then (sometimes) through a non-volatile fatty or mineral oil. The fairly pure NH_3 gas is then led into distilled water until the required concentration (up to about 36 per cent. NH_3 is attainable) is obtained.

The charcoal filters are revivified from time to time by ignition in closed retorts.

The following table gives the specific gravity of aqueous ammoniacal solutions of various strengths (after Lunge and Wiernek):—

SPECIFIC GRAVITY OF AQUEOUS SOLUTIONS OF AMMONIA AT 15° C.

Specific Gravity.	NH_3 per Cent.	1 Litre Contains g. HN_3.	Specific Gravity.	NH_3 per Cent.	1 Litre Contains g. NH_3.
1.000	0.00	0.0	0.936	16.82	157.4
0.998	0.45	4.5	0.934	17.42	162.7
0.996	0.91	9.1	0.932	18.03	168.1
0.994	1.37	13.6	0.930	18.64	173.4
0.992	1.84	18.2	0.928	19.25	178.6
0.990	2.31	22.9	0.926	19.87	184.2
0.988	2.80	27.7	0.924	20.49	189.3
0.986	3.30	32.5	0.922	21.12	194.7
0.984	3.80	37.4	0.920	21.75	200.1
0.982	4.30	42.2	0.918	22.39	205.6
0.978	5.30	51.8	0.916	23.03	210.9
0.974	6.30	61.4	0.914	23.68	216.3
0.972	6.80	66.1	0.912	24.33	221.9
0.968	7.82	75.7	0.910	24.99	227.4
0.966	8.33	80.5	0.908	25.65	232.9
0.964	8.84	85.2	0.906	26.31	238.3
0.962	9.35	89.9	0.904	26.98	243.9
0.960	9.91	95.1	0.902	27.65	249.4
0.958	10.47	100.3	0.900	28.33	255.0
0.956	11.03	105.4	0.898	29.01	260.5
0.954	11.60	110.7	0.896	29.69	266.0
0.952	12.17	115.9	0.894	30.37	271.5
0.950	12.74	121.0	0.892	31.05	277.0
0.948	13.31	126.2	0.890	31.75	282.6
0.946	13.88	131.3	0.888	32.50	288.6
0.944	14.46	136.5	0.886	33.25	294.6
0.942	15.04	141.7	0.884	34.10	301.4
0.940	15.63	146.9	0.882	34.95	308.3
0.938	16.22	152.1	...	...	...

TECHNICAL AMMONIUM SALTS

Ammonium Sulphate, $(NH_4)_2SO_4$.—Manufacture and properties are described on p. 82 *et seq.* See also under **Manures, Martin's** "Industrial Chemistry," Vol. II.

Heat of formation is given by :—

NH_3 (gas) + $\frac{1}{2}H_2SO_4$ = $\frac{1}{2}(NH_4)_2SO_4$ (solid) + 25.5 Cal. = 1,594,100 B.T.U. per eq. ton S./A.

Ammonium Chloride (Sal Ammoniac), NH_4Cl.—Prepared by passing vapours of ammonium into HCl, or *vice versa.* Sal ammoniac is the sublimed chloride. It is somewhat expensive to sublime owing to the difficulty of obtaining suitable vessels into which to sublime the chloride. Cheap earthenware is often used, which, however, can only be used once, as it has to be broken to remove the sublimate of sal ammoniac.

The substance is used in galvanising, in soldering, in galvanic cells, in the manufacture of colours, in calico-printing, in pharmacy.

Heat of formation is given by :—

NH_3 (gas) + HCl (gas) = NH_4Cl (solid) + 41.9 kal. = 2,619,300 B.T.U. per eq. ton S./A.

Ammonium Carbonate.—Commercial ammonium carbonate is usually a mixture of ammonium bicarbonate, NH_4HCO_3, and ammonium carbonate, $NH_4O.CO.NH_2$.

The ammonia content of the mixture varies between 25-58 per cent., the usual percentage of ammonia being 31 per cent.

It is most easily made by direct combination of ammonia, carbon dioxide, and water vapour, the substance being condensed on water-cooled surfaces of aluminium. The substance is volatile, and should at once be packed into air-tight vessels in order to avoid loss.

It is used in the manufacture of baking powders, dyeing, in extracting colours from lichens, in caramel making, smelling salts, etc. Also as a general detergent, and for removing grease from fabrics.

Ammonium Nitrate, NH_4NO_3.—See **Martin's** "Industrial Chemistry," Vol. II., for manufacture and properties. Its main use is for safety explosives, sporting powders, fireworks, etc. Much is used for the preparation of nitrous oxide, "laughing gas."

Heat of formation is given by :—

NH_3 (gas) + HNO_3 = NH_4NO_3 (solid) + 27 kal. = 1,687,900 B.T.U. per eq. ton S./A.

Ammonium Perchlorate, NH_4ClO_4, is prepared by the double decomposition of $NaClO_4$ and NH_4Cl.

For Patents bearing on its manufacture see **Alvisi,** D.R.P., 103,993; **Miolati,** D.R.P., 112,682; see also **Witt,** *Chem. Ztg.*, 1910, p. 634.

The salt is obtaining extended use as an explosive and oxidising agent, soluble in 5 parts cold water, insoluble in alcohol, sp. gr. 1.89.

Ammonium Phosphate, $(NH_4)_2HNO_4$, is made by neutralising phosphoric acid with ammonia, evaporating, and crystallising. Used in the manufacture of sugar (Lagrange process), and in the impregnation of matches.

Ammonium Persulphate, $(NH_4)_2S_2O_8$, prepared by electrolysing an acid saturated solution of ammonium sulphate at 7° C., using 5 volts in a specially designed apparatus. Much used as an oxidising agent in dyeing and photography; see the D.R.P., 155,805, 170,311, and 173,977; also Marshall, *Trans. Chem. Soc.*, 1891, p. 777.

Ammonium Thiosulphate, $(NH_4)_2S_2O_3$, prepared by double decomposition, thus—$2NHCl + Na_2S_2O_3 = (NH_4)_2S_2O_3 + 2NaCl$. Used in photography.

Ammonium Acetate, $NH_4OOC.CH_3$, prepared by neutralising acetic acid with ammonia, and used for making acetamide.

Ammonium Fluoride, NH_4F, made by neutralising HF with NH_3, is used for etching glass, decomposing minerals containing rare earths, for the manufacture of incandescent mantles, for preparing antimony fluoride and other technically important metallic fluorides, and also, to some extent, in dyeing.

Ammonium Sulphocyanide, NH_4CNS, occurs in gas liquor, and is often prepared by adding flowers of sulphur to an ammoniacal coal-gas washer, ammonium polysulphide being formed and cyanogen absorbed. It can also be made from CS_2 and NH_3:—$CS_2 + 2NH_3 = NH_4CNS + H_2S$. Used in photography, calico-printing, and dyeing.

Ammonium Chlorate, NH_4ClO_3, used in fireworks and explosives.

Ammonium Bromide, NH_4Br, used in pharmacy and photography.

Ammonium Oleate is used in ammonia soaps.

Dry Ammonia, $CaCl_2 8NH_3$, containing 60 per cent. of NH_3, is made by direct combination, and has in Germany a market as a portable and compact form of ammonia.

"**Solid Ammonia**" is manufactured by the Chemische Fabrik Betterhausen, Marquart & Schultz, by adding to a mixture of 3-5 parts of sodium stearate (dissolved in 10 parts of aqueous ammonia or 80 per cent. spirits of wine) about 85 or 90 parts of ammonia solutions containing 25-33 per cent. of ammonia. The mixture has a consistency nearly equal to that of solid paraffin. When heated or left exposed to air it gives up the whole of its ammonia, leaving behind the solid sodium stearate.

Large quantities of ammoniated superphosphates are made in the United States of America, containing up to 6 per cent. NH_3, and made by treating the superphosphate with ammonia or merely by mixing in ammonium sulphate into the superphosphate. Products are used for manurial purposes.

Anhydrous Ammonia (Liquid Ammonia) is simply the purified NH_3 gas liquefied under pressure and filled into steel cylinders.

It is the most suitable and efficient working substance for refrigerating machines, and is used in some wool-washing institutions for cleaning purposes, the substance being an excellent solvent.

The liquid boils at 34° C. At 15° C. the liquid has a vapour tension of 6 atmospheres. As a liquid it has a very large coefficient of expansion, and a specified weight at 15° C. of 0.614, at 60° C. of 0.540.

In order to manufacture the substance, excess of lime is added to crude gas-water so as to fix all the CO_2 and H_2S, as well as to set free all the ammonia from the "fixed" ammonium salts. The liquid is then distilled in a special column apparatus, somewhat similar to that described on p. 82 for ammonium sulphate, but somewhat more complicated, the large masses of mud-like lime requiring special vessels for mixing, depositing the precipitated mud, and for boiling out the ammonia. The NH_3 gas emerging from the columns is cooled and then made to traverse a number of vessels containing milk of lime, whereby the last traces of CO_2, H_2S, etc., are removed. The vapour then passes through a layer of paraffin oil, which retains tar and pyridine, etc. Then the gas passes through charcoal filters to remove the last traces of tarry matters, etc. The dry and pure vapour next passes to the pumps, where it is liquefied under 8 atmospheres' pressure. The operation is usually carried out in two stages, the gas heated in the first compression pump being cooled by water cooling before being passed into the next pump where the final liquefication takes place.

The liquid is stored in steel cylinders, usually made to hold either 20 kg. or 50 kg. of liquid ammonia. The cylinders should be tested every four years at 30 atmospheres' pressure. For each 1 kg. liquid ammonia there is allowed a volume of 1.86 l. The liquid ammonia, on evaporation, should not leave behind more than 0.1 per cent. residue, consisting of water, machine oil, pyridine, etc.

SYNTHETIC AMMONIA

LITERATURE

F. HABER and LE ROSSIGNOL.—"Technical Preparation of Ammonia from its Elements," *Zeit. für Elektrochemie*, 1913, **19**, 53, 72. (There is a good abstract of this paper in the *Journ. Soc. Chem. Ind.*, 1913.) *Ber.*, 1907, **40**, 2144; *Zeit. Elek.*, 1908, **14**, 181, 513.

HABER and VAN OORDT.—*Zeit. Anorg. Chem.*, 1905, **43**, 111; 1905, **44**, 341.

BERNTHSEN.—"Eighth International Congress of Applied Chemistry." New York, 1912. Abstract, *Journ. Soc. Chem. Ind.*, 1912, **31**, 982.

DONATH and INDRA.—"Die Oxydation des Ammoniaks zu Saltpetersäure und Saltpetriger Säure," pp. 54-67. Stuttgart, 1913.

LUNGE.—"Coal Tar and Ammonia." Fourth Edition, pt. 2, p. 815.

KNOX.—"The Fixation of Atmospheric Nitrogen." 1914.

NORTON.—"Utilisation of Atmospheric Nitrogen." 1912.

CARO.—"Die Stickstroffrage in Deutschland." 1908. *Zeitsch. angew. Chem.*, 1910, **23**, 2412.

SERPEK.—English Patent, 13,086, 1910.

J. W. RICHARDS.—*Trans. Amer. Electrochem. Soc.*, 1913, **23**, 351.

FRAENKEL.—*Zeitsch. Elektrochem.*, 1913, **19**, 362.

S. A. TUCKER.—*Journ. Soc. Chem. Ind.*, 1913, **32**, 1143; *Journ. Ind. and Eng. Chem.*, 1913, **5**, 191.

Also references and patents given in the text.

SEVERAL methods of making atmospheric nitrogen unite to form ammonia have been proposed, and have been commercially successful. In the course of a few years large amounts of synthetic ammonia will be on the market produced by one or other of these methods.

(1) AMMONIA BY DIRECT UNION OF NITROGEN AND HYDROGEN BY MEANS OF A CATALYST

LITERATURE

F. HABER and R. LE ROSSIGNOL.—"Technical Preparation of Ammonia from its Elements," *Zeitschrift für Elektrochemie*, 1913, **19**, 53-72.

Under certain conditions nitrogen directly unites with hydrogen to form ammonia according to the equation:—

$$\underset{\text{1 vol.}}{N_2} + \underset{\text{3 vols.}}{3H_2} \rightleftarrows \underset{\text{2 vols.}}{2NH_3}.$$

Heat evolved according to the equation:—

$$N + 3H = NH_3\,(\text{gas}) + 11{,}900 \text{ calories} = 727{,}000 \text{ B.T.U. per eq. ton S./A.}$$

This equation is reversible, depending upon the equilibrium represented by the expression:—

$$K = \frac{P_{NH_3}}{P_{N_2}^{\frac{1}{2}}\,P_{H_2}^{\frac{3}{2}}},$$

where P_{NH_3}, P_{N_2}, P_{H_2} represent the partial pressures of the respective gases, NH_3, N_2, H_2, and K is the reaction constant.

It will be noticed that 4 vols. of the mixture of nitrogen and hydrogen produce 2 vols. of ammonia, and consequently, as in the case of all gaseous reactions where the products of interaction occupy a smaller volume than the original components, and increase of pressure favours the formation of the products possessing the least volume. Haber and his co-workers found that by employing a very high pressure, about 200 atmospheres, and a temperature of between 500°-700° C., and by passing the mixture of gases over a catalyst, such as osmium or uranium, the combination of nitrogen with hydrogen proceeded so favourably that from 3-12 per cent. of ammonia was formed in the reacting gases.

It will be seen from the above equation that large amounts of energy are not required for the production of ammonia from its elements, and that, therefore, the manufacture of synthetic ammonia need not be confined to districts where large amounts of cheap water-power are available, as is the case with the electrical production of nitric acid, cyanamide, etc.

The Badische Anilin und Soda Fabrik have now erected works at Oppan, near Ludwigshafen, for the large scale preparation of synthetic ammonia by this process.

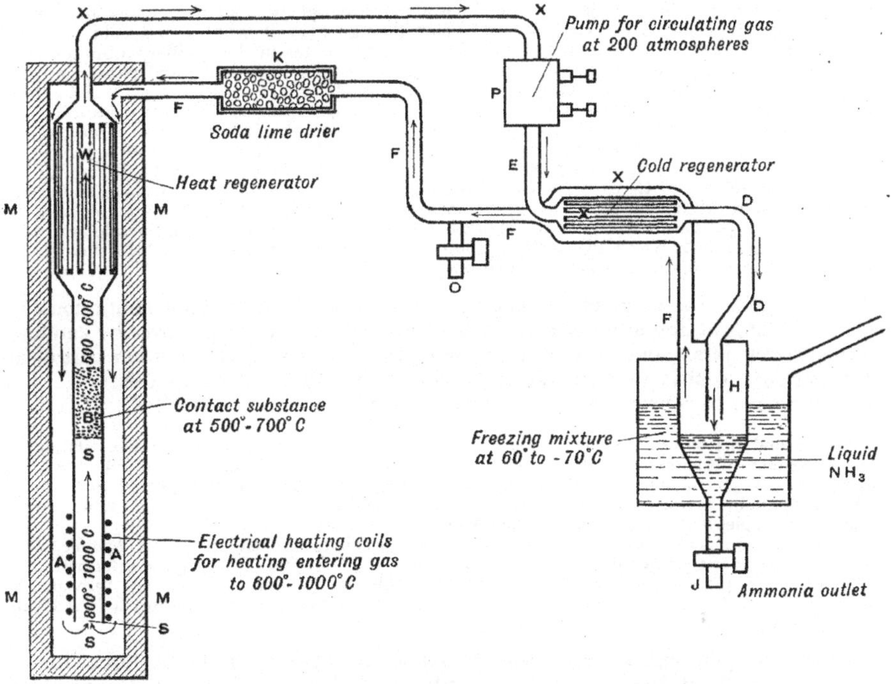

FIG. 46.—The Haber and Le Rossignol Process for Synthetic Ammonia.

Although the details of the plant employed on the large scale have not been published, the details of the experimental plant employed by Haber and Le Rossignol have been described in the above publication.

A diagrammatic sketch of Haber and Le Rossignol's experimental apparatus is given in Fig. 46.

Through the tube F a mixture of 1 vol. nitrogen and 3 vols. of hydrogen under a pressure of 200 atmospheres enters the strong steel vessel MM. After passing over the outer surface of a number of capillary metallic tubes W—which serve as a heat interchanger and regenerator as we will presently explain—the gas passes down the tube as shown, over the surface of an electrical heating coil AA, where the temperature of the gas is increased to 800°-1,000° C., then back up an interior iron tube SS, over the layer of catalytic substance B, thence through a number of capillary tubes W, out through the tube XX, thence through the compressing pump P, working at 200 atmospheres' pressure, thence out through the tube E, through the set of capillary tubes X, and so into the vessel H, which is surrounded by a freezing mixture of solid CO_2 and ether at a temperature of −60° or −70°, which causes the ammonia in the gas to separate in the liquid state, whence it can be drawn off by the cock J. From H the gases pass away by the tube F, over the exterior surface of the system of capillary tubes X, thence after passing over a soda-lime drier K, the gas enters M as previously described.

The mode of action of the apparatus is as follows:—The cold gas entering MM by the pipe F is heated by passing over the bundle of capillary tubes W conveying the hot gas away from the contact substance B. Thus the entering gas is, by the time it has left W, preheated to a temperature of 400°-500° C., and in so doing has abstracted practically all the excess heat from the hot gas passing away from B, so that this latter, by the time it reaches the pipe XX, is practically at atmospheric temperature, while at the same time the entering gas, by the time it reaches the heating coil A, is already at a high temperature, so that practically no loss of heat occurs. For this reason W is called the "heat regenerator." The hot gas thus entering the tube SS is further heated in its passage by the electrical heating coils surrounding the end of the tube SS to a temperature of 800°-1,000° C. The hot gas then passes into the contact substance B, which is maintained by the hot gas at a temperature of 500°-700° C. Here the formation of ammonia takes place, 3-7 per cent. of the entering nitrogen and hydrogen escaping as NH_3, along with excess of uncombined nitrogen and hydrogen. The hot mixture of gases from B then streams through the series of fine capillary tubes W, and in so doing gives up practically all its heat to the cold entering stream of gas coming into the apparatus at F. The gaseous mixture, now cooled to ordinary temperatures, passes away through the pipe XX into the pump P, working at 200 atmospheres, and then passes through the series of metallic capillary tubes X. While passing through these it meets with a cold stream of gas at −60° C. coming from the vessel H. Consequently the gas in the capillary tubes XX parts with its heat to the cold gas coming from H, being itself chilled in so doing, and passes out of X through the pipe DD into H at a temperature only slightly above that of the cold gas escaping from H. The cold gas passing from H up F is heated almost up to atmospheric temperature by the capillary tubes X, and thus passes away through the drier K and enters MM at F at a temperature only very slightly below atmospheric. For this reason the tubes X are called the "cold regenerator."

The gas entering H contains 3-7 per cent. of NH_3, and this condenses in a liquid form at the low temperature (−60° C. to −70° C.) prevailing therein, owing to the surrounding freezing mixture of solid CO_2 and ether. This ammonia can be drawn off in a liquid form by the tap J, or, if required in a gaseous form, can be so obtained by opening the outlet valve to a suitable extent.

As the ammonia is withdrawn a fresh supply of nitrogen and hydrogen is added through the valve O, so that the operation is a continuous one.

In practice, the very serious engineering operations of working continuously a plant with gas at 200 atmospheres, without leakage, has been got over by carefully turned screw joints, one part of an angle of 16° C. screwing into another part of an angle of 20°, so that perfectly gas-tight connections were made in this way. For special details of construction the original memoirs should be consulted.

Haber and his co-workers have made numerous experiments on the most suitable catalysts to use as contact substances, describing the results of experiments with cerium and allied metals, manganese, tungsten, uranium, ruthenium, and osmium. The best catalyst proved to be finely divided osmium, but as this substance is limited in quantities and very expensive, it was found that uranium (pieces the size of a pin's head) also acted efficiently.

Thus, in one series of experiments, commercial uranium, broken up with a hammer, was used in a column 4-5 mm. diameter and 3-3.5 cm. long. At 600° C. a vigorous formation of ammonia took place. At 190 atmospheres and with the gas mixture passing through the apparatus at 20 l. per second (measured at atmospheric pressure and temperature) it issued with an ammonia content of 5.8 per cent. by volume.

The nitrogen can be obtained from the atmosphere, either by liquefying it and fractionally distilling it with a Linde or Claude machine, as described in **Martin's** "Industrial Chemistry," Vol. II., or simply by passing air over heated copper. The hydrogen can be obtained industrially by any of the methods discussed in **Martin's** "Industrial Chemistry," Vol. II.

In 1913 ammonia in the form of commercial 25 per cent. (NH_3) ammonium sulphate possessed a value of 4.75d. per lb. (89 Pf. per kilo), while the nitrogen and hydrogen contained therein may be valued at 1.07d. per lb. (2½ Pf. and 17½ Pf. for the H and N respectively, in 1 kilo of ammonia).

It has been stated that the total costs of manufacturing ammonium sulphate from this synthetic ammonia only amounts to £2. 6s. 6d. per ton, and so the process appears to be the one which has a greater chance of success as regards low cost than any other method yet proposed, especially as the plant can be erected anywhere (see **Martin's** "Industrial Chemistry," Vol. II.).

Other Processes for Making Synthetic Ammonia from Atmospheric Nitrogen and Hydrogen.—De Lambilly (German Patents, 74,274 and 78,573) passes a mixture of N_2, steam, H_2 and CO over Pt (or other catalyst) at 80°-130° C., when ammonium formate is formed ($N_2 + 3H_2 + 2CO + 2H_2O = H.COONH_4$). With CO_2 at 40°-60° C. ammonium bicarbonate is formed ($N_2 + 3H_2 + 2CO + 2H_2O = 2NH_4HCO_3$).

Schlutius (English Patent, 2,200, 1902) passes Dowson gas (39 per cent. CO, 4 per cent. CO_2, 43 per cent. N_2, 14 per cent. H_2) and steam over Pt in the presence of a silent electrical discharge. Below 80° NH_3 is produced, above 80° C. ammonium formate.

The reader may also see the patents:—Young, English Patent, 1,700, 1880; Hooper, U.S. Patent, 791,194; Cassel, German Patent, 175,480; Gorianoff, French Patent, 368,585; see also Davies, *Zeit. physical. Chem.*, **64**, 657; Briner and Mettler, *C. R.*, **144**, 694; Donath and Indra (*loc. cit.*, pp. 54-67) and Lunge, "Coal Tar and Ammonia," Fourth Edition, pt. 2, p. 815, give details of other processes.

(2) **Ammonia from Cyanamide.**—Cyanamide is made by causing atmospheric nitrogen to directly unite with calcium carbide, as described in **Martin's** "Industrial Chemistry," Vol. II. Ammonia is next made from the cyanamide by passing superheated steam over it, when the following change takes place:—

$$CaNCN + H_2O = CaCO_3 + 2NH_3.$$

The operation may be carried out as indicated in Fig. 47.

Superheated steam is led in through the pipe A into the chamber B filled with the cyanamide on trays as indicated.

Ammonia is here generated, which may be directly drawn off or else led into acid and fixed.

The crude calcium cyanamide contains, it will be remembered, much carbon in the form of **graphite**, the substance as put on the market having the approximate composition CaNCN + C.

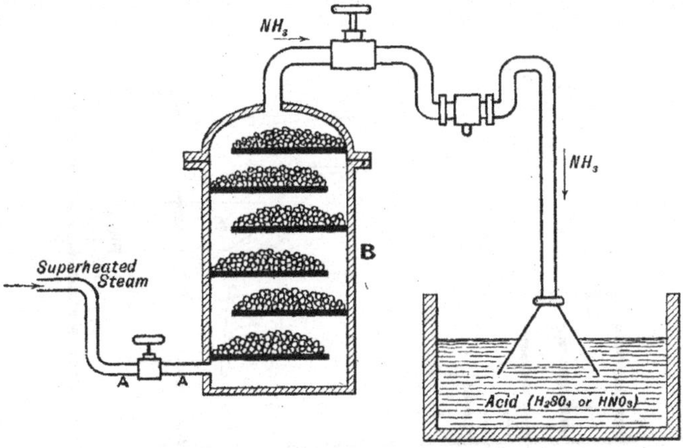

FIG. 47.—Ammonia from Cyanamide.

After treating with steam, as above described, we have the residue of $CaCO_3 + C$ left. This may be either worked for graphite (by dissolving the $CaCO_3$ in acid which leaves the C as graphite unaffected), or the residue may be returned to the lime furnaces, and be there calcined for the production of CaO, the lime thus produced containing one-third the proper amount of free carbon necessary for the production of calcium carbide when returned to the calcium carbide factory ($CaCO_3 + C = CaO + C + CO_2$ and $CaO + 3C = CaC_2 + CO$).

According to J. Louis Foucar, since cyanamide has to be made from calcium carbide and nitrogen, and the calcium carbide in its turn from lime and a plentiful supply of electrical energy, the ammonia made by this process could not, theoretically, be produced as cheaply as the direct synthetic ammonia prepared by direct union of nitrogen and hydrogen. Foucar (private communication) worked out the costs of manufacture of ammonium sulphate from cyanamide as follows:—

Cost of carbide	£3 5 0 per ton.	
Cost of nitrogen	0 2 0	
Cost of steam	0 0 6	
Cost of sulphuric acid	0 16 0	
		£4 3 6
Less value of graphite		0 5 6
Total net cost for materials		£3 18 0 per ton.
Cost for powder (cyanamide only)		0 7 6
Labour (cyanamide, steaming, etc.), power, repairs, depreciation, rent, taxes, amortisation, salaries of staff, and other charges		0 7 6
Total cost of manufacture		£4 13 0 per ton

The selling price of ammonium sulphate in England in 1912 was £14 per ton, and the total cost of manufacture from gas-liquor was £3 per ton.

(3) **Ammonia from Nitrides.**—A great many proposals have been made to use nitrides, either directly as fertilisers, or to produce ammonia therefrom by the action of superheated steam.

One process actually in use is the **Serpek Process**,[1] in which **aluminium nitride**, AlN, is used.

Bauxite (a naturally occurring impure hydrated alumina) is heated with coal in an atmosphere of nitrogen at a temperature of 1,700°-1,800° C. in a specially designed electric furnace, when N is absorbed, thus:—

$$Al_2O_3 + 3C + N_2 = 2AlN + 3CO.$$

The absorption of N begins at 1,100° C.; at 1,500° C. the absorption is rapid, the velocity increasing up to 1,800°-1,850° C., where almost violent absorption takes place, nearly chemically pure AlN being produced. Above 2,000° the nitride decomposes.

The reaction is strongly endothermic, Fraenkel calculating the heat absorbed in the above reaction to be −243,000 calories, and Richards, −213,220 calories.

According to Richards (*loc. cit.*) two superimposed cylinders, A and B, Fig. 48, which rotate in opposite directions, are used. Powdered bauxite is sent into A at C

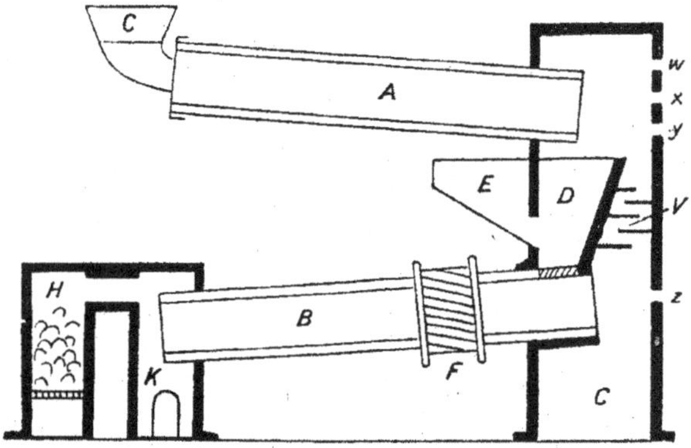

FIG. 48.—The Serpek Process.

and is calcined in its descent by the hot gases from the reaction going on in the lower cylinder B, and also by the combustion of the CO gas coming from the producer K and evolved in the reaction, this combustion being carried on in the side furnace C, air being admitted by the flues *w*, *x*, *y*, *z*, and the baffle plates V remove dust from the ascending gases. As the result of this the highly heated bauxite falls into a hopper D, and is there mixed with the requisite amount of carbon by means of the side hopper E. The bauxite and carbon fall into the lower rotating cylinder B (made of iron lined with compressed aluminium nitride), and thence passes into the electric resistance furnace F (made of a series of bars of compressed carbon and AlN, crossing the furnace diametrically or embedded longitudinally in the lining of the furnace), where the mixture is heated to 1,800°-1,900° C., and there meets the nitrogen from the producer-gas plant H (evolving 70 per cent. N_2 + 30 per cent. CO at 400° C.) placed at the lower end of the apparatus.

The nitrogen is absorbed, and the resulting aluminium nitride, in the form of a grey powder, passes on to an air-tight chamber K at the bottom of the apparatus.

The silicious impurities in the charge are largely volatilised out; consequently, using crude bauxite, the mass contains 26 per cent. N, but using pure alumina, Al_2O_3, 34 per cent. N can be obtained, corresponding to pure AlN.

The resulting aluminium nitride, AlN, is next decomposed by caustic soda to form ammonia and sodium aluminate:—

$$AlN + 3NaOH = NH_3 + Na_3AlO_3.$$

From the sodium aluminate thus produced, pure alumina can be obtained in the manner described in **Martin's** "Industrial Chemistry," Vol. II.

According to the Badische Anilin u. Soda Fabrik Patents (German Patents, 235,300, 235,765, 235,766, 235,868, 236,395) the elimination of ammonia by alkalies is facilitated by the addition of NaCl and other soluble salts. The decomposition of the AlN can be effected by acids, thus:—$2AlN + H_2SO_4 + 6H_2O = 2Al(OH)_3 + (NH_4)_2SO_4$. These patents work out methods for recovering the $Al(OH)_3$ in a pure state.

The same firm in the German Patent, 243,839, states that the formation of nitride from alumina and coal is greatly accelerated by the addition of 5-10 per cent. of the oxides of certain elements, which themselves yield stable nitrides, among which SiO_2 and the oxides of Ti, Zr, V, Be, Mo, Ur, Ce, Cr, also silicates, vanadates, etc. Iron oxide does not influence the catalytic action of these oxides.

Aluminium nitride, as prepared by Serpek, is of value both for its combined nitrogen and also for the production of pure alumina for the manufacture of metallic aluminium. Without the sale afforded by this alumina it is doubtful whether the process would be economical, as the alumina would have to be used over and over again. As it stands, however, the alumina produced as a by-product finds a ready sale, and the ammonia simultaneously produced is also a valuable product.

The process is being worked by the Société Générale des Nitrates in France.

OTHER NITRIDES

Various suggestions have been made to use **boron nitride**, BN, but the high temperature required for its formation and the volatility of the resulting boric acid when the nitride is decomposed by superheated steam for ammonia ($2BN + 3H_2O = 2B(OH)_3 + 2NH_3$) have formed insuperable difficulties.

Calcium nitride, Ca_3N_2, containing 18.9 per cent. of N, and **lithium nitride**, Li_3N, containing 38.8 per cent. N, are easily prepared from the elements and easily decompose, giving ammonia. Magnesium nitride, Mg_3N_2, containing 27.7 per cent. N, is likewise easily prepared by heating magnesium in nitrogen gas. When heated in hydrogen it is converted into ammonia and hydride ($Mg_3N_2 + 6H_2 = 2NH_3 + 3MgH_2$).

However, none of the processes suggested to employ these reactions as technical sources of ammonia appear to be successful.

The attempts to utilise **titanium nitride** by heating it in a mixture of nitrogen and hydrogen have also been unsuccessful.

Silicon nitride, Si_3N_4, containing 42.5 per cent. N, has been suggested for direct use as a fertiliser, as giving results possibly superior to ammonia and cyanamide. The Badische Anilin und Soda Fabrik (English Patent, 16,368 of 1910) have patented processes for fixing atmospheric nitrogen by means of silicon nitride. They point out that in preparing silicon-nitrogen compounds from silicic acid, carbon and nitrogen, this operation formerly had to be conducted in the electric furnace owing to the high temperature necessary. They find, however, that it is possible to conduct the reaction at a low temperature if oxides, hydroxides, or salts of metallic elements are added to the mixture of silicic acid and carbon. They also suggest the employment of silicious materials such as the silicates of iron, aluminium, calcium, etc. Instead of pure nitrogen they employ gaseous mixtures containing N.

They give as an example 75 kilos of finely divided silica mixed with 25 kilos of powdered wood charcoal, and heated in a stream of nitrogen for ten to twelve hours at 1,300°-1,400° C., allowing the product to cool in N gas. When the product is treated with saturated or superheated steam, ammonia is produced.

It is doubtful, however, whether any of these processes are likely to be a commercial success, in view of the successful production of synthetic ammonia directly from the elements.

However, the reader may consult the following patents:—Basset, English Patent, 4,338, 1897; Lyons and Broadwell, U.S. Patent, 816,928; Borchers and Beck, German Patent, 196,323; Roth, German Patent, 197,393; Kaiser, English Patent, 26,803, 1905; Wilson, English Patent, 21,755, 1895; Mehner, English Patents, 12,471, 1895, 2,654, 1897, 28,667, 1903.

CHAPTER IX

Manufacture of Sulphur Dioxide and Sulphites

CHAPTER IX

MANUFACTURE OF SULPHUR DIOXIDE AND SULPHITES

By Geoffrey Martin, D.Sc.

Manufacture of Sulphur Dioxide and Sulphurous Acid

The burner gases (see **Martin's** "Industrial Chemistry," Vol. II.) from pyrites, zinc blende, or spent oxide from gas works are sometimes worked for the manufacture of **anhydrous sulphur dioxide** (SO_2). The usual process is that of **Hanisch** and **Schröder** (see Fig. 49).

The gases from the roasting furnaces, which must contain at least 4 per cent. SO_2 by volume, otherwise too much water is required for complete condensation of the SO_2, enter through A, pass under the leaden pans B B and heat them, then pass into the tower C C, packed

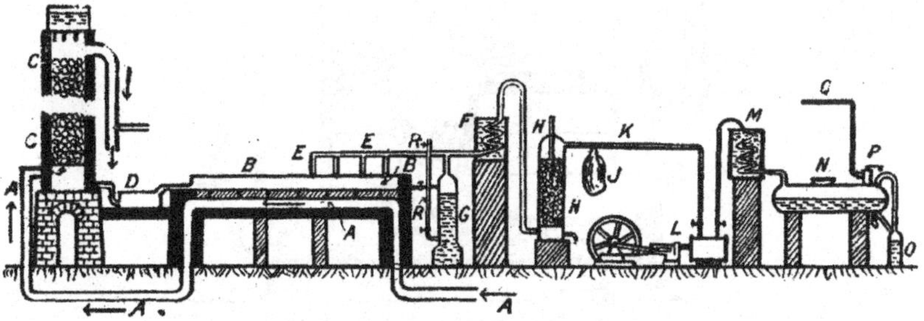

Fig. 49.—Hanisch and Schröder's Process for the Manufacture of Anhydrous Sulphur Dioxide.

with coke, down which cold water trickles. This absorbs practically all the SO_2, which thus collects as a solution at the bottom of the tower, and from there runs into the preliminary heater D, which is composed of thin lead plates, round which hot waste liquids (from a later stage of the operations) circulate and give up their heat usefully. The SO_2 solution leaves D at a temperature of about 85° C., and then runs over the leaden pans B B, where the heat of the flue gases causes it to boil, thus expelling SO_2 through the pipe E E to the cooling worm F, where it is cooled, and most of the moisture accompanying it runs away into G. The SO_2 gas is then dried in a tower H H, packed with coke down which a stream of concentrated sulphuric acid trickles, and then passes a taffeta bag J (which regulates the pressure) along the pipe K into the bronze pump L, where, under a pressure of 2½ atmospheres, it is liquefied, cooled in M, run into the cast-iron boiler N, whence it can be drawn off as required. The moisture saturated with SO_2, which collects in G (a small tower packed with lead wire), is boiled by live steam entering through R R. The SO_2 is thus boiled out and recovered, whereas the hot liquid so obtained is used for heating the preheater D.

The anhydrous liquid SO_2 so obtained is sold commercially of 99.8 per cent. purity.

For many purposes, *e.g.*, for bleaching sugar, cellulose, etc., also for treating wood pulp for making into paper (see **Martin's** "Industrial Chemistry," Vol. I.), there is required a more or less dilute **solution of sulphur dioxide in water**.

This is easily obtained by bubbling the gas, rich in SO_2 either from pyrites burners or from other suitable burners, through water (after first washing with water).

In order to obtain a strong solution both gas and water must be well cooled.

For many purposes it is essential that the sulphurous acid (SO_2 solution in water) should be free from sulphuric acid. (*E.g.*, the presence of sulphuric acid in the bisulphite used for treating wood pulp is very injurious to the fibre.)

Hence, as pyrites burner gas nearly invariably contains SO_3, and the plant for burning the pyrites must be close to the place of consumption of the aqueous solution, many manufacturers find it cheaper to make their own sulphurous acid and sulphite by **burning sulphur** in a suitable burner, and leading the gas into water.

There are two general types of furnace in which sulphur is burned: of these one intentionally volatilises the sulphur and ensures complete combustion of the gas, the other burns some of the sulphur, and condenses that portion which is volatilised but not burned. Of these there are again sub-types.

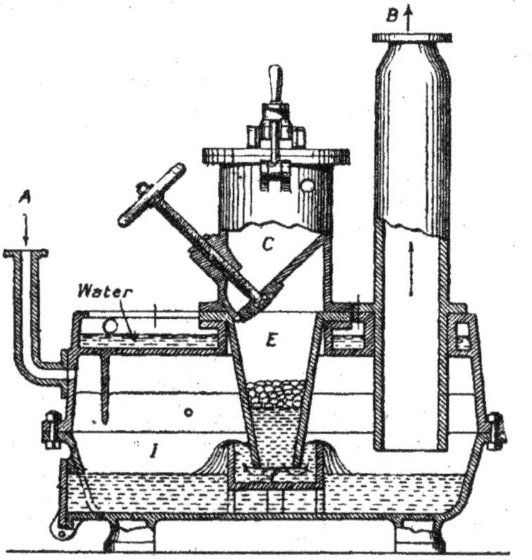

FIG. 50.—Sachsenberg Sulphur Burner.

Of the first type the most noteworthy is the **Tomblee and Paull.** It consists of a horizontal, cylindrical, lined iron shell 8 ft. long and 3 ft. in diameter, with conical ends, making one revolution in two minutes. At one end is a hopper and worm-feed with sliding dampers, at the other a rectangular cast-iron box of 37 cub. ft. capacity, provided with sliding damper and vertical uptake, which leads to a brick dust-catcher. The sulphur is fed in lumps which melt just before dropping into the body of the burner, and a complete liquid coating is formed on the inside of the shell as it revolves. The combustion box and vertical uptake complete the combustion. The furnace, box, and uptake are all lined with refractory material. Using 98 per cent. Louisiana sulphur, it is only necessary to clean out the dross which collects once in two months. The appliance will burn 6,000 lbs. of S per day.

Hand-fed burners of the first type are now completely obsolete, and will not be described; some of them employ secondary air to ensure the complete combustion of the sulphur. The writer has found this very satisfactory if such air be preheated.

Of the second type of furnace may be mentioned the **Sachsenberg burner** (Fig. 50). It is particularly suitable for the production of SO_2 in **small** quantities, and which is required to be as free as possible from SO_3, as is required for bleaching, in the manufacture of glue, in the refining of sugar, etc.

In starting up sulphur is placed in and ignited by inserting a red-hot iron bar through the hole in the body (which, when working, is closed by a plug). The hopper C and funnel E are filled with sulphur, which melts and forms a liquid seal in F. Air is drawn through A, and the SO_2 passes out through B. The upper portions of the body and funnel are water-cooled. Compressed air may be introduced into B to burn any sulphur sublimed there. The slight coating of sulphur on the pipes, etc., is found to effectually preserve them.

A burner of this type, measuring only 8 ft. by 3 ft., will burn no less than one ton of sulphur per twenty-four hours. See also p. 106 for a complete plant for SO_2 manufacture.

A more modern type of plant is that patented and produced by Mr H. B. P. Humphries, of Queen Anne's Chambers, London, S.W.

Fig. 51 shows a furnace of medium size, while a complete installation is outlined in Fig. 52.

The plant may be worked under considerable gas pressure, and arrangement is made for

FIG. 51.—Humphries' Sulphur Burner.

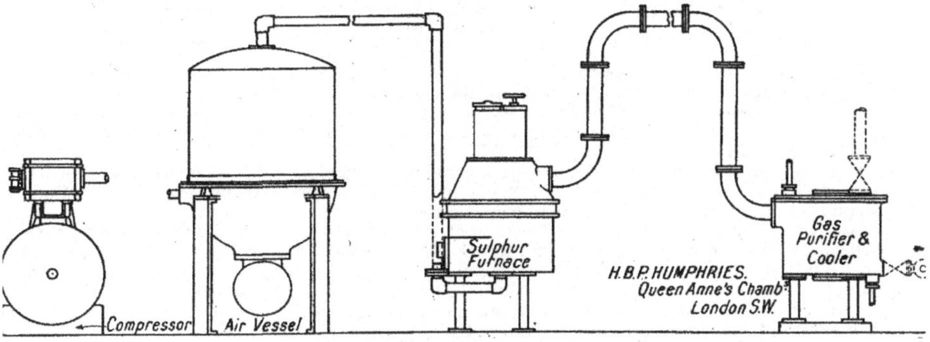

FIG. 52.—Humphries' Plant for making Sulphur Dioxide.

replenishing the sulphur without any disturbance whatever of the working. Special attention is given to the construction of the joints and to facilities for cleaning out. The latter is seldom necessary—once every few weeks, or maybe months—depending on the work required from the plant. For some purposes, such as sugar defecation, a feature is incorporated which diminishes the content of SO_3 in the gases to a minimum, conformably with the most recent recommendations of sugar experts.

Installations of single units and of batteries are in use throughout the world. The smallest size burns 2½ lbs. sulphur per hour, while a large size burns 50 lbs. per hour.

The sulphur furnace, cooler, and system in general are covered by patents.

Properties and Uses of Sulphur Dioxide and Sulphurous Acid.—Colourless gas, with a pungent suffocating smell, incombustible, and a non-supporter of combustion. Sp. gr. = 2.26 (air = 1).

By cooling or compression it may be condensed to a colourless mobile liquid, boiling at 10° C. under atmospheric pressure. The pressures exerted by liquid SO_2 at various temperatures are:—

Temperature	−10° C.	+0° C.	+10° C.	+20° C.	+30° C.	+40° C.	+50° C.
Pressure (in atmosphere)	0	0.53	1.26	2.24	3.51	5.15	7.18

Critical temperature = 156° C.; critical pressure = 78.9 atmospheres. The latent heat of evaporation is large, being 93.4 calories per kilo at −10° C.

Liquid SO_2 does not act upon iron, and is therefore put up in steel cylinders, being used in refrigerating machines. The gas is also used in bleaching, and for antiseptic purposes.

Its chief advantages for this purpose are its fairly high latent heat and pungent smell, the latter giving timely warning of any escape.

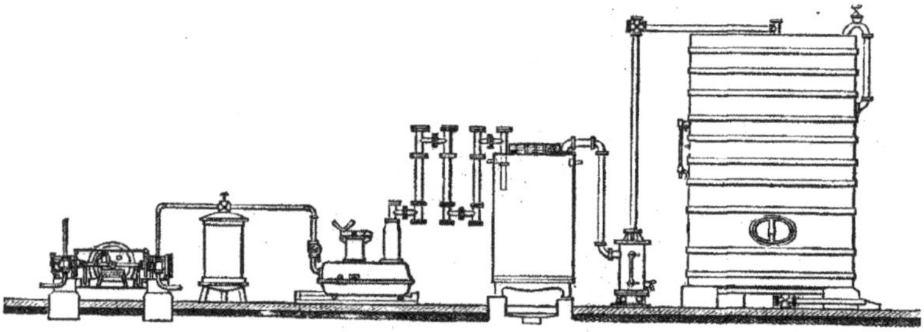

FIG. 53.—Complete Plant for Making Solution of Sulphur Dioxide.

SO_2 is very soluble in water (water at 0° C. dissolves 79.8 vols. of SO_2, at 20° 39 vols., at 40° C. 19 vols.); considerable heat is evolved in the process of solution, and the liquid behaves as if it were a solution of sulphurous acid (H_2SO_3).

It corrodes many metals, such as wrought iron and zinc, decomposes carbonate, forms salts with most metals, and generally behaves like an acid.

Uses.—The gas or its aqueous solution is mainly used for **bleaching** and **antiseptic** purposes, in **sugar purification**, and in making large numbers of organic chemicals.

The most important use is undoubtedly in paper making, where vast quantities are used in the form of calcium bisulphite. Large quantities are also used in the Hargreaves' saltcake process (see **Martin's** "Industrial Chemistry," Vol. II.).

Much **sodium bisulphite** is used under the name of "antichlor," in order to withdraw the last traces of chlorine from goods bleached with chlorine. Solid sodium sulphite ($Na_2SO_3 + H_2O$) and potassium metasulphite ($K_2S_2O_5$) are both made on the large scale.

Analysis.—Sulphites or other combinations of sulphurous acid are readily recognised by their smell on acidification with sulphuric acid, by the bleaching properties of the resulting gas, by its discoloration of solutions of permanganate and iodine, and its reduction to H_2S in presence of nascent hydrogen (zinc or aluminium). Sulphur dioxide may be estimated by standard iodine titration, by oxidation to sulphuric acid and precipitation as barium chloride or titration with standard alkali in the case of its aqueous solution, and in the absence of other acids.

The air of towns near metallurgical works, smelters, or calciners usually contains sulphur dioxide, which slowly oxidises with the formation of sulphuric acid or possibly of ammonium sulphate. If it exceeds a certain amount ($\frac{1}{3000}$ volume), serious damage will be done to vegetation, which cannot seem to become accustomed to it.

CHAPTER X

Acetylene

CHAPTER X
ACETYLENE

By Frank B. Gatehouse, F.C.S.

Acetylene, C_2H_2, is the simplest member of the acetylene series of unsaturated hydrocarbons having the general formula C_nH_{2n-2}.

Acetylene is a gas at ordinary temperature and pressure: it occurs in small quantities in coal gas and is produced during the incomplete combustion of methane, ethyl alcohol, coal gas, and other substances.

It may be synthesised from its elements by causing an electric arc to pass between carbon poles in a current of hydrogen.

Acetylene may also be produced by various other methods which have, however, become only of academic interest since the discovery of calcium carbide by Wilson in Canada, and Moissan in France, in 1892. The growth of the carbide and acetylene industry since that date has been phenomenal, and its history forms one of the romances of science.

For commercial and industrial purposes acetylene is now produced entirely by the interaction of calcium carbide and water:—

$$CaC_2 + 2H_2O = C_2H_2 + Ca(OH)_2.$$

Pure acetylene is said to be almost odourless, the well-known unpleasant odour usually associated with this gas being due to impurities, chiefly phosphorous compounds (see below).

Acetylene has a theoretical specific gravity at 0° C. of 0.9039 (air = 1) which may be compared with Leduc's experimental value of 0.9056. The critical temperature of acetylene is 37° C. and the critical pressure nearly 68 atmospheres. At lower temperatures acetylene is liquefied at a less pressure, but owing to the fact that it is liable to decompose or dissociate into its elements with explosive violence even under a pressure of 2 atmospheres, the manufacture, storage and use of liquefied acetylene is prohibited by law, but this does not apply to "dissolved acetylene," which see.

Acetylene is readily soluble in water, 100 vols. of water dissolving about 95 vols. of C_2H_2 at ordinary temperature and pressure; it is less soluble in brine, lime-water or calcium chloride solution, which facts are occasionally taken advantage of in practice. For details as to solubility in acetone see under "**Dissolved Acetylene,**" p. 113.

Acetylene has an ignition temperature in air of 406°-440° C. (Dixon and Coward). Mixtures of acetylene and air have a very wide range of explosibility ranging from 3.35 to 52.3 per cent. of acetylene by volume. Acetylene is readily polymerised by heating, the chief product being benzene (C_6H_6).

The heat of combustion of acetylene has been determined by various experimenters. It may be stated to be 363 large calories per gramme-molecule. Numerous attempts to determine the temperature of the acetylene flame have also been made. Under ordinary circumstances it may be taken at about 2,500° C. in air, and 4,000° in oxygen (see **Welding**).

Calcium Carbide is manufactured in the electric furnace, the raw materials being lime and coke or anthracite coal.

The principal impurities to be guarded against in the raw materials are phosphorus, sulphur, silica, magnesia, iron, and alumina, which either give rise to impurities in the evolved acetylene or difficulties in the manufacture of the carbide.

The **storage of carbide** in this country is governed by an Order in Council under the Petroleum Acts, which together with local bye-laws and regulations should be studied by intending users of acetylene.

The Analysis of Calcium Carbide and Acetylene.—In the space at our disposal it is impossible to more than briefly indicate the impurities that should be estimated for technical purposes. Full details as to the methods to be employed will be found in the volumes of *The Acetylene Lighting and Welding Journal* mentioned under "Literature," and in Lunge and Keene's "Technical Methods of Analysis."

Apart from the gas-yield of carbide, the question resolves itself into an estimation of the impurities present in the acetylene evolved under proper conditions.

The gas-yield can best be estimated by decomposing a known weight of carbide in a suitable generator, and passing the gas through a standardised meter. All the water employed should be first saturated with acetylene and the mean of several results taken. Owing to the difficulty of sampling, as large a quantity of carbide as possible should be employed; for this reason estimations carried out under ordinary laboratory conditions cannot be relied upon.

The regulations of the British Acetylene and Welding Association which govern the purchase and sale of carbide in this country enjoin that the carbide when broken to standard size (1-2½ in. or larger) shall yield not less than 4.8 cub. ft. per lb. at standard temperature (60° F.) and pressure (30 in.).

A latitude of 5 per cent. shall be allowed for analysis. For other important details the regulations, which may be obtained from the office of the Association, 103 and 104 Cheapside, London, E.C., should be consulted.

The only impurity specified is phosphoretted hydrogen, which must not exceed 0.05 per cent. by volume, with a latitude of 0.01 per cent.

It is desirable also to test for sulphur compounds and ammonia in the acetylene and also to ascertain the temperature and speed of reaction between the carbide and water. The former should not exceed 100° C.

Acetylene for Domestic and Industrial Purposes

Until recent years acetylene was used almost solely as an illuminant for the lighting of private houses, churches, institutions, also less extensively for street and public lighting.

Owing to the great improvements that have been made in the construction of Bunsen burners, it may now be quite satisfactorily employed for heating and cooking purposes, and also in conjunction with specially prepared mantles for incandescent lighting.

Acetylene is used largely for **marine signals, beacons,** and **buoys,** and for temporary lighting with portable generators, "**flare lamps,**" etc. Acetylene motor, motor cycle and bicycle lamps are too well known to need any description; it is obvious, however, that owing to contingent circumstances acetylene generated and used under such conditions is placed at a great disadvantage.

Acetylene generators as ordinarily employed for domestic and industrial purposes may be divided into two classes: (1) Non-automatic, and (2) Automatic. These again may be divided into carbide-to-water and water-to-carbide generators.

Briefly, in the non-automatic type a quantity of carbide is decomposed at a stated time, and the whole of the gas liberated is collected in a gas-holder for future consumption; in the automatic type, the carbide is decomposed in limited and stated quantities for immediate consumption, the reaction between the carbide and water being regulated by the consumption of the gas.

A complete generating plant consists of a generator, washer, condenser, gas holder, and purifier, with the necessary connections, valves, etc.

Our illustrations depict (Fig. 54) a non-automatic carbide-to-water plant, (Fig. 55) an automatic water-to-carbide generator, and (Fig. 56) the well-known "Carbic" plant in which specially prepared carbide cakes are used.

These plants are typical of the classes mentioned, and are made by some of the best known acetylene engineering firms in the United Kingdom. The diagrams are self-explanatory. We are indebted to "The Rosco" Acetylene Co., of 130 York Street, Belfast, "The Thorn and Hoddle" Acetylene Co. Ltd., Westminster, and "Carbic" Ltd., London, for permission to use these illustrations.

The purifying materials employed, chiefly for removal of the phosphorus compounds, are based either on bleaching powder or chromic acid. The former substance, however, has several drawbacks in use, and is being replaced by the more satisfactory chromic acid compounds.

One of the best known purifiers is "**heratol**" which is stated to consist of an acid solution of chromic acid absorbed in "Kieselguhr." A more recently introduced material is known as "**Catalysol**" (French Patent, 434,137, 12th September 1911, and French Patent, 440,207, 15th February 1912, P. J. Granjon).

This consists of kieselguhr, 350 parts, soaked with a solution of ferrous chloride (40° Bé.), 640 parts; mercuric chloride, 8 parts; precipitated manganous acid, 2 parts. The mixture is exposed to the air until the ferrous salt is completely oxidised, when it forms a yellow powder which may be used until it becomes white. The oxidation and revivification is facilitated, it is claimed, by the incorporation of a small quantity of copper sulphate (or other copper salts) with the liquid or mass.

One of the chief advantages of this material is that when exhausted it may be revivified by exposure to air. Both in the case of heratol and catalysol there is a gradual change in colour as the material becomes exhausted, but the degree of purification of acetylene may always be easily ascertained by placing a strip of filter paper soaked in nitrate of silver solution in a stream of the gas as it issues from a burner. If efficiently purified the test paper will not darken.

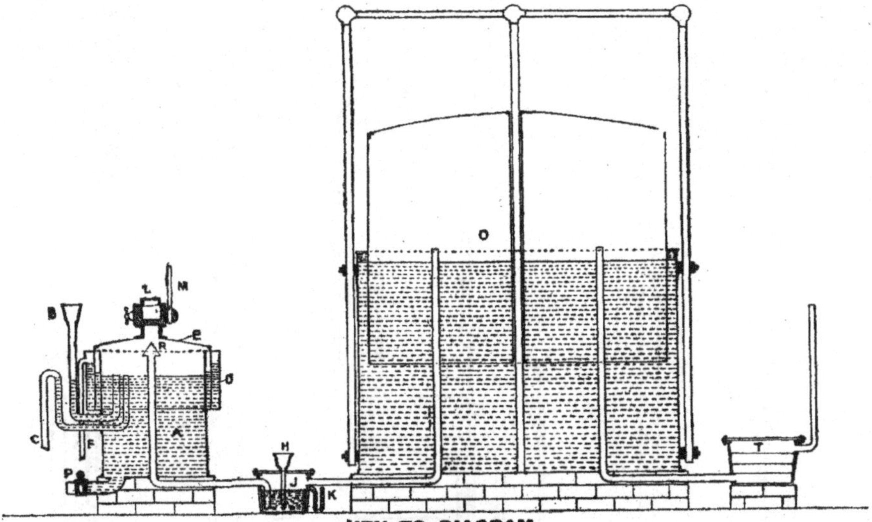

KEY TO DIAGRAM.

FIG. 54. —"Rosco" Patent Non-Automatic Acetylene Plant.

Key to Diagram.—(A) Generator. (B) Water Inlet to Generator. (C) Overflow. (D) Water Seal. (F) Water Overflow of Seal. (L) Patent Carbide Feed-Valve, which consists of a Hollow Plug Tap with Handle (M); when the Handle (M) is turned upwards the hollow of the Tap can be filled with Carbide; on giving the Tap a third of a turn the Hollow of the Plug is cut off from the outer air and is opened to the Generating Tank so that the Carbide contained in it is discharged over a Distributor (R) unto the Tray (N) which is in the water in the Generating Tank. (P) Cock for removal of Sludge. (J) Scrubber and Seal Pot through which Gas passes to Holder. (H) Water inlet to Scrubber. (K) Overflow of Scrubber. (O) Gasholder. (T) Dry Purifier.

We have mentioned that acetylene may be used with incandescent mantles with special burners, but the acetylene must be rigorously purified and used at a

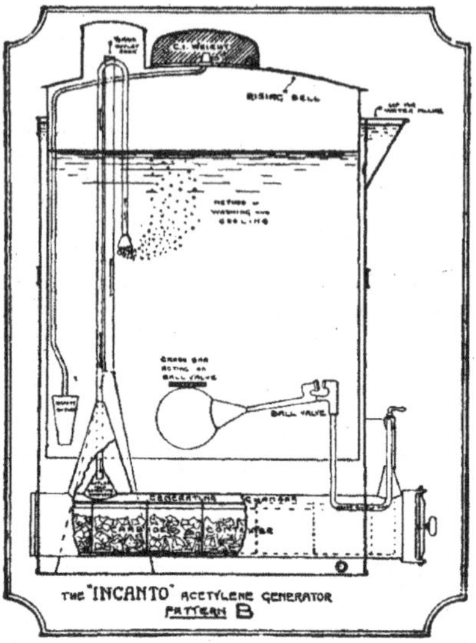
THE "INCANTO" ACETYLENE GENERATOR
PATTERN B

FIG. 55.

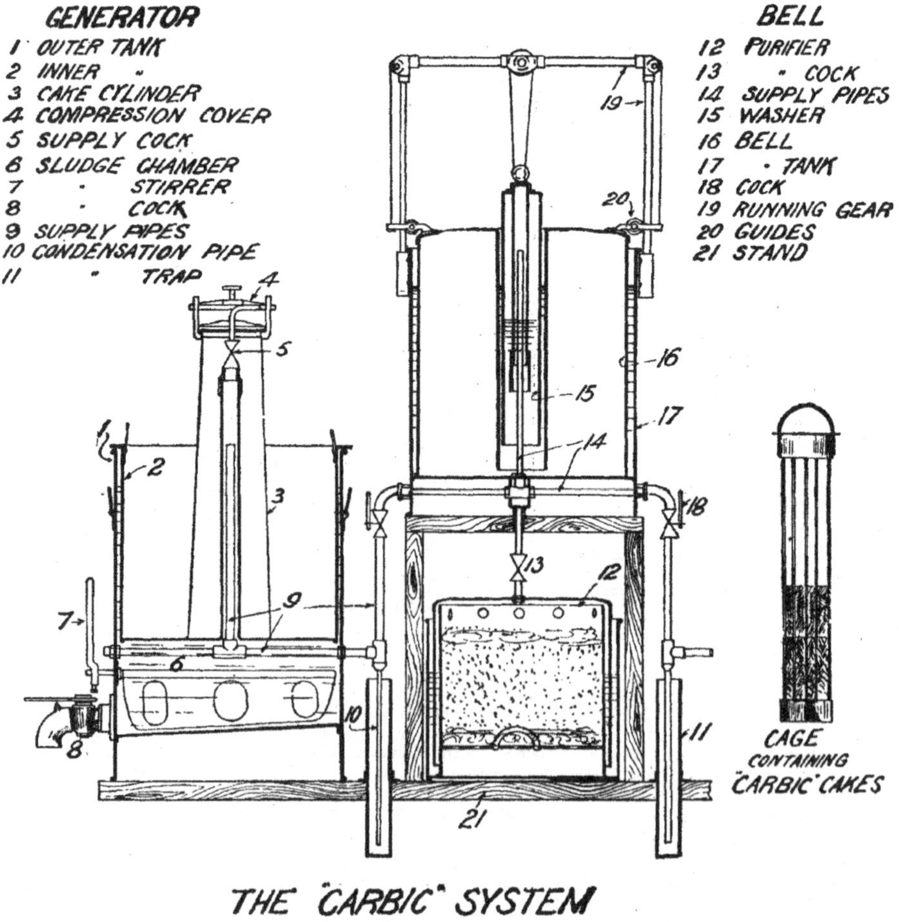
GENERATOR
1 OUTER TANK
2 INNER "
3 CAKE CYLINDER
4 COMPRESSION COVER
5 SUPPLY COCK
6 SLUDGE CHAMBER
7 " STIRRER
8 " COCK
9 SUPPLY PIPES
10 CONDENSATION PIPE
11 " TRAP
BELL
12 PURIFIER
13 " COCK
14 SUPPLY PIPES
15 WASHER
16 BELL
17 " TANK
18 COCK
19 RUNNING GEAR
20 GUIDES
21 STAND
CAGE CONTAINING "CARBIC" CAKES
THE "CARBIC" SYSTEM

FIG. 56.

higher pressure than when ordinary atmospheric burners are employed. This is not altogether a disadvantage, as the gas may then be more satisfactorily employed for heating and cooking purposes.

In this country, however, the use of atmospheric burners such as the well-known "Elta," "Luta," "Cyka," and the "Roni," made by Messrs George Bray & Co., Ltd., of Leeds, are almost exclusively employed. All acetylene generators and apparatus should conform to the regulations of the British Acetylene and Welding Association.

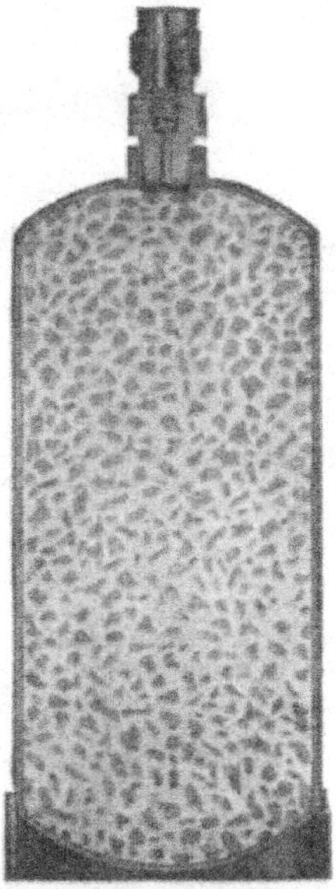

FIG. 57.—Acetylene in Acetone. (Acetylene Illuminating Co.)

Oxy-Acetylene Welding.—For the autogenous welding of metals and alloys, also for cutting purposes, the oxy-acetylene welding process is making phenomenal progress, and bids fair to eclipse all other welding methods.

For the low-pressure system of welding, the acetylene is obtained from a suitable generator as described above, in the high-pressure system the gas is obtained from a cylinder of dissolved acetylene, under pressure; in both cases the oxygen is generally also obtained from a cylinder. Special welding and cutting blowpipes are employed, also suitable welding rods, and in some cases fluxes are required. The process is of the widest application and is largely used both for constructional and repair work, but serious work should not be attempted or permitted without expert tuition and guidance. Welding schools have been formed at many of the chief industrial centres, and courses for workmen, works managers, and engineers are held regularly.

For further details of the process reference should be made to the works mentioned under "Literature." Welding blowpipes, valves, manometers, etc., may be obtained from the British Oxygen Co., Ltd., Westminster, and most acetylene engineering firms.

Dissolved Acetylene.—As mentioned previously acetylene is readily soluble in many organic liquids, notably acetone. Advantage is taken of the fact, and at the same time the difficulties and dangers connected with the manufacture of the compressed gas overcome, in the manufacture of what has become known as "Dissolved Acetylene." The largest, and until recently the only, firm manufacturing dissolved acetylene in the country is the Acetylene Illuminating Co., of Lambeth Road, S.W. According to their process the washed, cooled, and purified acetylene is pumped under pressure in cylinders apparently filled with a suitable porous material saturated with acetone.

FIG. 58. The "Roni" Single Orifice Burner.

FIG. 59. "Elta" Burner for Acetylene. Can be turned low like coal gas burners without carbonising.

Acetylene Burners. (Bray.)

1 vol. acetone takes up, under 12 atmos. pressure, about 100 vols. of acetylene.

Dissolved acetylene is employed chiefly for high-pressure oxy-acetylene welding, motor car and railway carriage lighting, and for marine signals and beacons.

For railway carriage lighting, etc., carburetted acetylene has been employed. This consists of a mixture of acetylene and a hydrocarbon vapour such as benzol, but its employment has not become very extensive.

Acetylene in the Chemical Industry.—The applications of acetylene for lighting, heating, and especially welding purposes are many and varied, notably in the construction of chemical plant and vessels, drums, etc., but apart from these applications acetylene is finding an ever-widening field of usefulness as a chemical agent as the basis of new and interesting compounds.

Calcium Carbide, the "raw material" for the production of acetylene, has itself become an important substance in the chemical industry as the starting-point in the manufacture of various compounds of nitrogen such as the cyanamides, cyanides, ammonia, etc.

8

The compounds of acetylene with chlorine are already well known. They take the form of more or less volatile liquids, and may replace the highly inflammable carbon disulphide, ether, benzene, etc., in the extraction of fats and oils, etc.

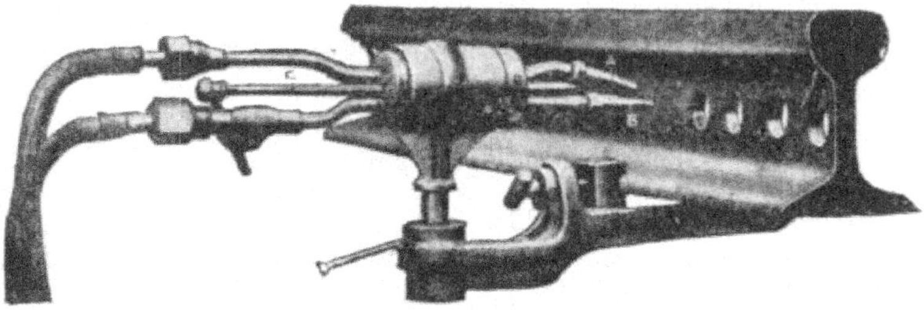

FIG. 60.—Oxy-Acetylene Flame used as Rail Piercer. A is Oxy-Acetylene Heating Jet, B is the Oxygen Jet, C is the Centering Bit; A and B Revolve. (Carbic Ltd., London.)

Acetylene tetrachloride or **tetrachlorethane** may be obtained by the direct union of acetylene and chlorine under the influence of light, but the reaction sometimes proceeds with explosive violence. It is manufactured according to one process by passing acetylene and chlorine alternately through a mixture of finely divided iron and chloride of sulphur. The tetrachloride is distilled off and the residual sulphur converted into the chloride for use again.

Among other processes enjoining the use of acetylene that have been suggested or patented may

FIG. 61.—Cutting Armour Plate 9 inches thick with Oxy-Acetylene Flame. (Supplied by Carbic Ltd., London.)

be mentioned the manufacture of **phenol** by a process devised by Berthelon, as well as the production of various sulphuric derivatives of acetylene, which apparently, however, have not been carried out on a commercial scale.

By heating a mixture of acetylene with either of the oxides of carbon, or by passing the oxide over heated carbide, free carbon is obtained in the **graphitic state.** These reactions are stated to be

CHAPTER XI

The Illuminating Gas Industry

CHAPTER XI
THE ILLUMINATING GAS INDUSTRY

BY ERNEST A. DANCASTER, B.SC.

THE illuminating gas industry originated in England with Murdoch in 1792. London streets were illuminated by gas in 1813, Paris in 1815, Berlin, 1826. The industry is to-day an enormous one. England still produces the most gas, distilling in 1909 nearly 16,000,000 tons of coal and producing 6,900 million cub. yds. of gas, representing 150 cub. yds. per head for the whole country and 29·2 cub. yds. for London. The introduction of Auer's incandescent mantle in 1885 opened a new epoch in the coal-gas industry, which was at one time seriously threatened by the electric light industry. More recently the introduction of carburetted water gas, either alone or more usually mixed with ordinary coal gas (say 10-20 per cent. carburetted water gas in the coal gas), represents a fresh development, since by the addition of this gas the heating and luminous value of the gas can be greatly increased, and be made independent of external disturbances.

The production of a luminous gas is now of less importance than formerly, and, indeed, with the general introduction of the incandescent mantle, may be dispensed with, the heavy luminous hydrocarbons merely serving as sources of heat (a cubic metre of coal gas should yield at least 5,000 calories). The essential modern improvement, then, consists of producing luminosity ***not*** by glowing carbon particles (*e.g.*, the cause of the luminosity of the candle, oil, and ordinary gas flame), but by the more efficient glowing incandescent mantle.

COAL GAS

General.—Coal when heated in a closed space in absence of air undergoes destructive distillation, with the production of a residue of (*a*) coke, (*b*) tar, (*c*) aqueous fluid, and (*d*) gas.

Besides coal, we find wood, peat, bituminous shales, brown coal, bones, certain oils, and residues left in the manufacture of beet sugar subjected to destructive distillation on the industrial scale with the object of obtaining valuable chemical products.

Coal is subjected to destructive distillation (*a*) in gasworks, (*b*) in coke ovens, but with very different objects.

In **gasworks** the main object is to produce **gas** for illuminating purposes, and the other products—such as coke, ammonia and tar—are by-products. In **coke ovens**, however, the main object is to produce **coke** for metallurgical operations, the gas, tar, and ammonia being here by-products. In gasworks the coal is heated in small closed earthenware retorts, whereas in coke ovens the coal is de-gassed in large fire-brick chambers. For details regarding Coke Ovens see **Martin's** "Industrial Chemistry," Vol. II.

In Germany the coke ovens are greatly developed, over two-thirds of the total yield of tar and benzol being obtained from this source, together with very large amounts of illuminating gas. In England and the United States, however, coke ovens are but slightly developed, almost all the illuminating coal gas being obtained from gasworks. See p. 122.

For the economical production of coal gas a special sort of coal—the so-called gas coal—must be used, which yields a large amount of gas of good calorific power.

Gas coal usually contains 4-5½ per cent. H and 5-16 per cent. O, but a mere chemical analysis yields little information as regards the gas value of a coal; different coals—even when analysing the same—often behave quite differently, and the suitability of a coal for gas production is best ascertained by actual trial on a large scale. In general 1 kg. of good gas coal (calorific value =7,500 calories) should yield 300-350 l. of gas (calorific value, 1,500-1,850 calories) representing 20-25 per cent. of the heating value of the coal. 100 kg. of good gas coal should yield 16-19 kg. gas=30-35 cub. m., 65-68 kg. coke, 5 kg. tar, 8 kg. ammoniacal water, and 3 kg. loss. Coal on storage undergoes rapid deterioration, a continual loss of gas taking place, as well as surface oxidation.

Manufacture of Coal Gas in Gasworks

The Retorts.—In gasworks coal gas is always prepared by the destructive distillation of coal in small closed fire-clay retorts, externally heated to from 950°-1,400° C.

Horizontal Retorts.—The most common form is of D-shaped cross section, but oval and circular retorts are also known. They are usually set horizontally in beds of from three to twelve, heated by the same furnace. The walls of the ovens support the ends of the retorts, and a number of transverse partitions built in the ovens support them at intervals along their length. In large works these retorts are usually about 20 ft. long, and are fitted with an iron mouthpiece and self-sealing lid at each end. In small works shorter retorts, which are closed at one end, are used.

A, Fig. 64, shows the retorts in cross section, and Fig. 62 represents a longitudinal section through the furnace.

The heating is now invariably done by gas firing in regenerative furnaces, the heat of the burnt gases being utilised in heating that of the entering air. By this means an enormous saving in the amount of coke used for firing has been effected—the old coke fires required for every 100 tons of coal distilled about 30-40 tons of coke burnt, against 12-15 tons now required. Consequently at the present time at least 80 per cent. of the produced coke is available for selling. Moreover since gas fires give a much greater heat (1,400° C. in the hottest part), a large yield of gas is simultaneously obtained.

The firing gas, principally CO, passes into C (Fig. 62) from the gas-generator B, and is here mixed with a stream of pre-heated air, and burns, encircling the retorts as it does so on every side; the hot burnt gases pass away through a system of brickwork channels, running parallel to which is a similar and separate system of channels conveying the entering air, which is thereby intensely heated before it enters the gas chamber. The exhausted gases finally pass away through D. Immediately under the generator B is a water-tank over the surface of which the entering air passes, so that it enters B charged with moisture, which has the double effect of cooling the gas-producer, thereby prolonging its life, and of producing a little water gas. The gas generated in the retorts passes away up the tubes *gg*. Usually the gas is sucked out of the retorts by the exhauster (p. 123), so that the gas inside is at a pressure slightly less than atmospheric pressure (the pressure must not be too low or air is sucked in). This is done in order to withdraw the heavy hydrocarbon gases as rapidly as possible from the heat, in order that they may not be decomposed. Nevertheless much gaseous decomposition occurs, and in consequence the inner walls of the retort become lined with "retort carbon." Much of this carbon goes over as a fine dust with the gases into the hydraulic main.

At the close of the operation, when all the gas has distilled over, the lid of the retort is loosened, the issuing gas carefully ignited, and the gas remaining in the retort allowed to burn quietly away. The coke is then raked out and rapidly quenched with water. The retorts are recharged as soon as possible after the withdrawal of the coke. The charging is often done by hand by means of long shovels ("scoop-chargers"), worked by three men. In larger installations, however, a mechanical charger is used, usually consisting of a travelling trolley running on rails up and down in front of the retorts and fitted with a "scoop-charger" and coal hopper.

In West's manual charging machine, worked by two men, there is a double scoop which is fed with coal from the hopper. When full the scoops are pushed into the retorts by a rod and reversed by means of a handle attached to the rod. The charger is then withdrawn. An apparatus for breaking up coal is also provided, with an elevator for transferring the broken coal to a hopper which supplies the hoppers of the chargers as required. A mechanical drawing apparatus is also used. This consists of a rake which is manipulated by means of a bar held in channel guides.

Modern machine stokers, *e.g.* Fiddes-Aldridge, charge and recharge at one stroke.

The retorts last for 1-3 years, being kept during the whole of that time at a bright red heat.

Inclined retorts set at an angle of 30°-36° have come into use during recent years. These are usually about 15 ft. long, and of D-shaped cross section. The coal is poured into the retorts from a hopper, and the coke is withdrawn by opening the door at the lower end, when it falls out of itself, thus greatly diminishing the manual labour of charging and discharging.

Vertical retorts have been tried, but until recently with little success. Now, however, there are on the market vertical retorts which are far superior to the horizontal retorts above described. It was formerly thought that the coal would swell up and prevent the gas escaping freely, so that the resulting gas would be largely decomposed, and be poor in heavy hydrocarbons. Trial, however, has shown that the exact reverse holds true, the gas being of excellent quality on account of the fact that the inner core of coal remains relatively cool almost to the last, and the gas, escaping rapidly through this core, comes out rich in heavy hydrocarbons. The coke produced is denser and harder than that of horizontal retorts, the tar much thinner, and the yield of ammonia 50 per cent. greater, while that of the cyanides is smaller. Moreover, much less space is taken up by the retorts, and charging and discharging are especially easy—the retorts being made slightly larger at the base, so that the coke readily falls out, while the charging occurs by means of overhead hoppers.

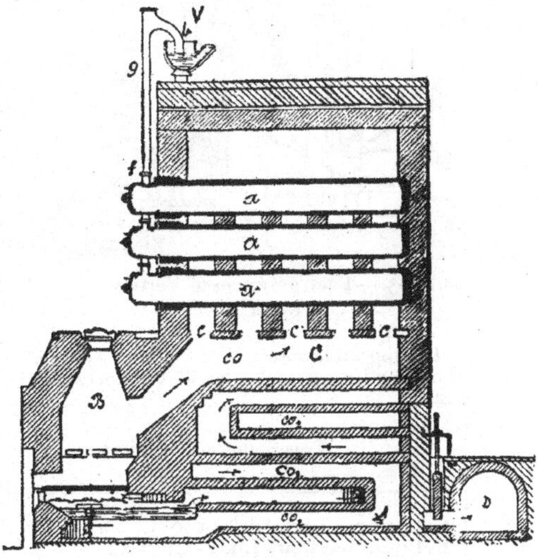

FIG. 62.—Longitudinal Section of Gas Retort Furnace.

Fig. 63 shows a vertical cross section of the new vertical retorts of Dr Bueb's system as erected by the Dessauer Gas Anstalt. R are the retorts, ten retorts in two rows, or eighteen retorts in three rows usually going to a furnace. G is the gas generator, which is fed with coke through *ef*. The producer gas, mixed with air, circulates as a very hot flame around the retorts in the flues *n*, the hot-burnt gases passing out through the regenerators *m*, where it gives up its heat to the entering air, and finally escapes through the smoke-flue *r*. The retorts are charged with coal by the hopper *a*, and the coke withdrawn below, being discharged through *c* into a trench *d*. The gas escapes through the hydraulic main V into the gas-main H, while the tar runs away by *k*.

Continuous carbonisers, such as *Woodall-Duckham* and *Glover-West*, have also been tried, and have proved very successful. Also **coking chambers**—similar to coke ovens—consisting of horizontal or inclined chambers, 2-3 m. high, and of a considerable length, holding 10-15 tons of coal, with a period of distillation of twenty-four hours (also continuous working ones), have been tried, and are fairly satisfactory.

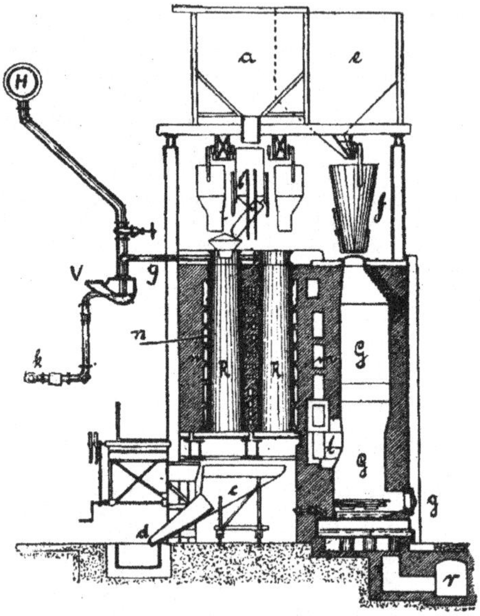

FIG. 63.—Bueb's System of Vertical Gas Retorts.

Course of the Distillation.—A temperature of about 1,000° C. is required for the de-gassing of coal; formerly a temperature of 960° C. was deemed sufficient, but now it is common to go up to 1,200°-1,400° C., and coal which has been heated to 1,500° C. still retains some H, N, and O. In general, the higher the temperature and the longer the period of heating the larger the volume of gas produced. The gas which comes off first at the lower temperatures is rich in heavy hydrocarbons and possesses much illuminating and heating power. The last portion of gas, obtained at the highest temperature, consists principally of hydrogen, and is of poor illuminating power. In other words, the lower the temperature of production the better the quality of gas but the smaller the quantity. A large volume of gas is invariably produced at the expense of its quality.

The following table shows this, the numbers referring to the volume of gas produced by distilling 100 kilos of English gas coal at various temperatures:—

Temperature of Distillation.	Volume of Gas Produced in Cubic Metres.	Candle-Power of Gas.
1. Dark red heat, 800°-1,000° C.	23.4	20.5
2. Bright red heat, 1,000°-1,100° C.	27.5	17.8
3. White heat, 1,200°-1,400° C.	34.0	15.6

The coal undergoes a loss of weight of 17-18 per cent. The following numbers show the variation in the composition of the gas evolved in successive stages of the distillation, the gas being collected in three successive lots of equal volumes:—

Stage of Distillation.	H.	CH_4.	CO.	Heavy Hydrocarbons.	CO_2.	Residue.
	Per Cent. (Vol.)	Per Cent. (Vol.)	Per Cent. (Vol.)	Per Cent. (Vol.)	Per Cent. (Vol.)	Per Cent. (Vol.)
1. First gas evolved	28.3	46.6	9.4	9.4	4.0	2.3
2. Intermediate gas	49.0	31.7	8.1	4.3	2.0	4.9
3. Last gas	55.3	27.2	8.8	1.7	1.8	5.2

The various portions of gas when mixed together analysed as H=45.2 per cent., CH_4=35 per cent., CO=8.6 per cent., heavy hydrocarbons=4.4 per cent., and residue=4.8 per cent.—all by volume. The original coal analysed as C=77.2 per cent., H=5 per cent., O+N=9.3 per cent., H_2O=2 per cent., ash=6.5 per cent., and yielded 17.7 per cent. by weight of gas=30.2 cub. m. per 100 kilos.

The N contained in the coal passes over partly as NH_3, partly as free N, a small fraction escaping as HCN, pyridine, quinoline, etc. The S (usually 0.5-1 per cent., sometimes 2 per cent.) escapes principally as H_2S and CS_2, but part remains behind in the coke.

Purification of the Crude Gas.—The gas coming from the retorts is very impure, containing tar, moisture, ammonia, sulphuretted hydrogen, hydrocyanic acid, naphthalene, etc., from which it must be freed before use.

The crude gas passes out of the retorts by the *ascension pipes g g* (Figs. 62 and 64), which lead it into the *hydraulic main* V (Fig. 64, shown in section in Figs. 62 and 63), where the bulk of the tar and moisture condense. The hydraulic main usually runs along the top of the retort bench, and the dip pipes which enter it are sealed by the liquid in the bottom of the main. This liquid consists of a layer of tar above which is a layer of ammoniacal liquor, both layers being deposited by the gas which has previously passed through. The tar and ammoniacal liquor are drawn off by a pipe so as to keep a constant level of both liquids. The hydraulic main is often divided into several sections, each of which is provided with an outlet pipe passing into a large main F (Fig. 64) known as the *foul main*. The water seal in the hydraulic main is sometimes replaced by mechanical devices known as anti-dips.

The gas is partly freed from ammonia, carbon dioxide, hydrogen sulphide, and

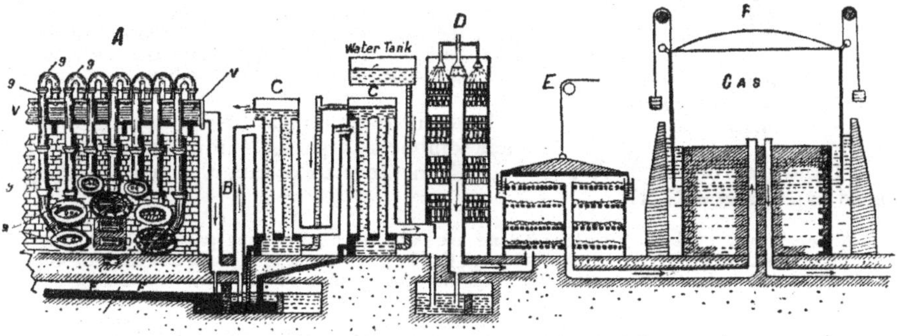

FIG. 64.—Section of Gas Works.

cyanogen by the liquor in the hydraulic main. It now passes into the air-condensers B (Fig. 64) where it is further freed from a quantity of tarry matter by deposition. In some works vertical condensers are used, but often the pipes, which are from 50-80 ft. long, are arranged in nearly horizontal rows which are exposed to the air. They are connected to a second row which passes underneath the first and so on; frequently there are about a dozen of these rows of pipes, one beneath the other, and cooled in summer by dripping water. Then the gas (especially in Continental works) is subjected to water cooling in C C, which is specially useful in hot weather.

Even after a very thorough cooling a portion of the tar remains suspended in the gas in the form of fine drops. Consequently in many works the gas passes into a "tar extractor," where it is made to pass through a series of perforated iron plates. Fig. 65 shows Pelouze's form, consisting of a hanging bell *a* suspended in water and made up of three layers of finely slit iron plates. The fine tarry drops in the gas condense to larger drops in passing through the slits, and so are separated. A regulator F raises or sinks the bell, and so regulates the rate at which the gas passes. Some ammoniacal liquid and practically all the tar is separated by this device.

Either before or after the tar extractor is placed the *exhauster*. This is really a pump used to draw the gas from the retorts and condensers and to force it through the washing and purifying apparatus. Sometimes fans or blowers are used for this purpose, when the gas has to be driven a considerable distance through small mains.

The gas now enters the *water scrubbers* D (Fig. 64). In some of the older works these take the form of tall towers packed with coke or wooden checker-work, down which water or weak liquor from the hydraulic main and condensers mixed with the washings from the last stage of the washing process is caused to trickle. Two at least of these scrubbers are used in series. More usually, however, some form of *washer scrubber* is used in which the gas is passed in fine streams through water or dilute liquor. Such scrubbers are far more efficient and occupy much less room.

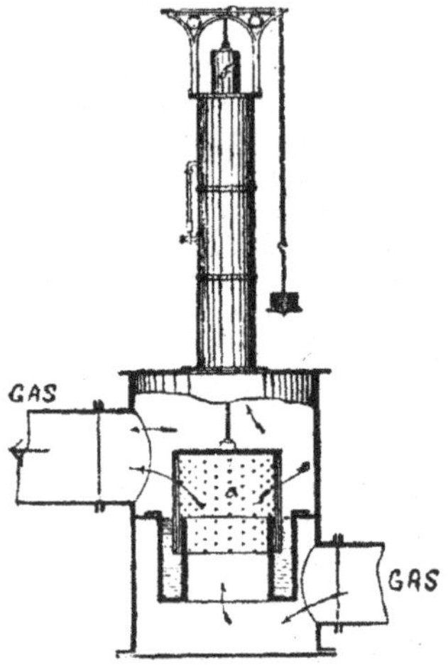

FIG. 65.—Pelouze's Tar Extractor.

In the "Standard" washer scrubber (Fig. 66) bundles of iron, or a combination of wood and iron B B, attached to a horizontal shaft F are slowly revolved, and thoroughly wetted by passing through water or weak liquor placed in the bottom of the washer. The gas is caused to pass through the wet bundles. The liquor passes through the apparatus in the opposite direction to the gas so that as the latter becomes more pure it meets a stronger solvent. Two such scrubbers in series are used, the first being fed with dilute liquor and the second with water. In the most modern form of this scrubber the bundles are made entirely of wood. The "New" washer scrubber has discs covered with bundles of Brazilian bass (*piaçaba*) fibre which press against the faces of the sections of the scrubber and the gas passes between the wet fibres. Between each pair of scrubbing chambers there is a still chamber.

In many works the gas before passing into the water scrubbers is passed through a scrubber half filled with tar oils of B.P. 250°-300° C. in order to extract the final residues of solid naphthalene, $C_{10}H_8$.

Although most of the naphthalene is condensed with the tar, yet some passes on as a vapour in the gas (1 cub. m. gas at 30° C. can hold 1 g. $C_{10}H_8$, at 0° C., 0.15 g.) and can lead to stopping up of the smaller section gas pipes. These tar oils take up over 16 per cent. naphthalene and are

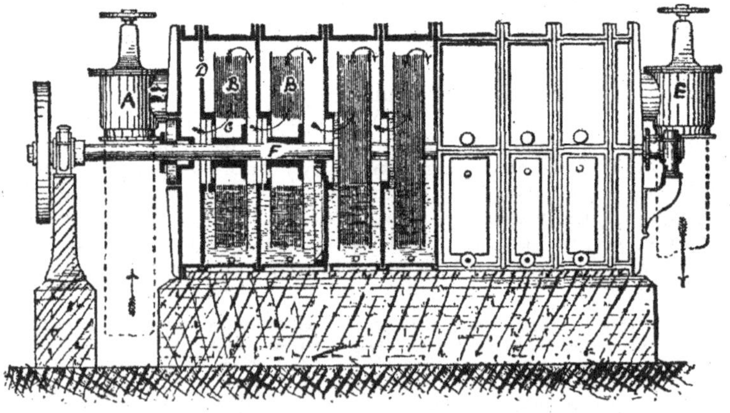

FIG. 66.—"Standard" Washer Scrubber.

then run off into the tar well. Since the tar oils could also extract benzol from the passing gas, they are charged with about 3 per cent. benzol to start with.

In the scrubbers practically all the ammonia, much CO_2, and some H_2S and HCN are removed. The resulting ammoniacal-gas water contains about 2 per cent. NH_3. The gas, however, still contains volatile sulphur compounds (principally H_2S with a little CS_2), HCN, and CO_2. To remove these the gas now passes through a series of *purifiers*, E (Fig. 64). These are usually rectangular iron vessels, the dimensions of which vary according to the amount of gas for which they are intended. The sides of the lid dip into a water seal. Inside these vessels are placed horizontal wooden (sometimes cast-iron) grids, placed one above the other. The purifying material is spread on these to a thickness of about 8 in. The gas enters the purifiers at the bottom and passes out at the top. In some Continental works a different method has been recently introduced in which the horizontal grids are replaced by composite grids which are so arranged as to cause the gas to traverse the purifying material chiefly in a horizontal direction.

A complete set of purifiers consists of eight vessels arranged in four pairs. As usually arranged the first pair are charged with slaked lime, the next two with hydrated ferric oxide mixed with sawdust or wood chips, and the third pair with sulphided lime (hydrated lime through which gas containing H_2S but no CO_2 has been passed). The last pair act as check vessels and contain lime and ferric oxide, or Weldon mud. Sometimes the iron oxide in the second pair of purifiers is replaced by Weldon mud which is essentially a hydrated manganese oxide and acts in the same manner as the ferric oxide. In this case the arrangement of the purifiers is sometimes altered, the second pair containing the sulphided lime and the third pair containing Weldon mud. The first pair always contains slaked lime.

At first the CO_2 and H_2S in the gas are absorbed by the lime in the first purifiers, but as the free lime gets used up the H_2S is liberated by further quantities of CO_2, so that ultimately all the H_2S passes on into the second pair of purifiers where it is absorbed by the moist ferric oxide (or Weldon mud), with the formation of iron (or manganese) sulphide. The gas then passes into the third pair of purifiers where the CS_2 is absorbed by the sulphided lime. It then passes through the last pair which remove any remaining traces of H_2S, and finally passes to the gas holder.

Frequently the sulphided lime vessels are left out, and sometimes the purification of the gas after leaving the washer scrubbers is limited to passage through hydrated ferric oxide only or through slaked lime alone.

This dry purification is not very perfect, since traces of sulphur (0.5-0.6 g. per cub. m.) and HCN still remain.

The chemical actions which take place are very various and complex. The slaked lime takes up CO_2: ($Ca(OH)_2 + CO_2 = CaCO_3 + H_2O$). It also takes up H_2S, forming $Ca(SH)_2$, which then reacts with any CS_2 forming the sulpho-carbonate, $CaCS_3$. CO_2 gradually decomposes the $Ca(SH)_2$, thus: $CO_2 + Ca(SH)_2 + H_2O = CaCO_3 + 2H_2S$. Moist iron oxide unites with the H_2S, thus: $2Fe(OH)_3 + 3H_2S = Fe_2S_3 + 6H_2O$, and the iron sulphide may be regenerated either by exposing to air in heaps, or by mixing a little air (1-1.5 per cent.) with the gas passing through the generator, when S is set free: $Fe_2S_3 + 3O = Fe_2O_3 + 3S$, and further H_2S is simultaneously oxidised in the presence of the moist iron hydroxide: $H_2S + O = H_2O + S$, which here acts as a catalytic agent. When the iron hydroxide contains 40-50 per cent. free sulphur it becomes useless for purifying and is removed.

Hydrocyanic acid, HCN, does not unite with the $Fe(OH)_3$ but with the iron sulphide, forming Prussian blue, $Fe_7(CN)_{18}$. A part, however, unites with the NH_3 and H_2S to form $NH_4S.CN$, ammonium sulphocyanide. Where it is important to remove all HCN, the gas is (previous to passing into the purifiers) passed through a 20-25 per cent. solution of $FeSO_4 + NH_3$ contained in a standard scrubber, when the insoluble ammonium ferrocyanide, $(NH_4)_3Fe_2.Fe(CN)_6$, is deposited, and one-third of the ammonia is fixed as ammonium sulphate: $FeSO_4 + H_2S + NH_3 = FeS + (NH_4)_2SO_4$.

Carburetted Coal Gas.—The luminosity of coal gas is sometimes increased by the addition of the vapour of very volatile hydrocarbons. This process is usually known as carburetting. The most common oil used for this purpose is "carburine" or light petroleum oil, but petroleum spirit is sometimes used, and also benzol.

The carburine is placed in a steel tank which is sunk in the ground. It is pumped as required from this tank to the evaporator which contains a number of tubes heated externally by means of steam. The carburine passes through these tubes and in doing so becomes vaporised. The vapour then passes on through a regulating valve and an injector into the gas main. The amount of vapour added to the gas is controlled by the regulating valve.

The following analyses show the difference effected in the composition of the gas by the purifying process:—

	Impure Gas from Retorts.	Purified Gas.
	Vol. per Cent.	Vol. per Cent.
H	46	49
CH_4	32	34
CO	8	8
Heavy hydrocarbons, C_2H_6, C_6H_6, etc.	4	4
NH_3	1	...
H_2S	1	trace
CO_2	4	1
Free N	4	4
	100	100

Mixed Coal Gas.—Of rapidly increasing importance (and which may in the near future lead to a great change in the whole gas industry) is the practice of adding carburetted water gas (see below, next section) to the extent of 10-20 per cent. to the coal gas. By this means the illuminating and heating power of the gas may be increased to almost any desired extent, while the gas company is rendered largely independent of the changing seasonal demand for gas, and of the partial failure of the coal-gas supply by an unforeseen accident.

The following analysis shows the composition of a purified coal gas mixed with 20 per cent. carburetted water gas:—

	H	CH_4	CO	Heavy Hydrocarbons, C_2H_4, C_6H_6	NH_3	H_2S	CO_2	N
Per cent. vol. =	50	27	14	4	...	...	1	4

Storage of the Gas is effected in large gas-holders (**F**, Fig. 64), "gasometers," of capacity 10,000-200,000 cub. m.—the largest in the world being in New York (500,000 cub. m.), which swim in deep water channels. The gas is distributed by means of a network of pipes, the main gas pipe being best of tarred cast iron, and 1 yd. in diameter. Through leakage in the distribution often 5 per cent. of gas is lost, though great improvements in this respect are now made. Usually 25 per cent. of the cost of the gas is due to the cost of the distributing pipes.

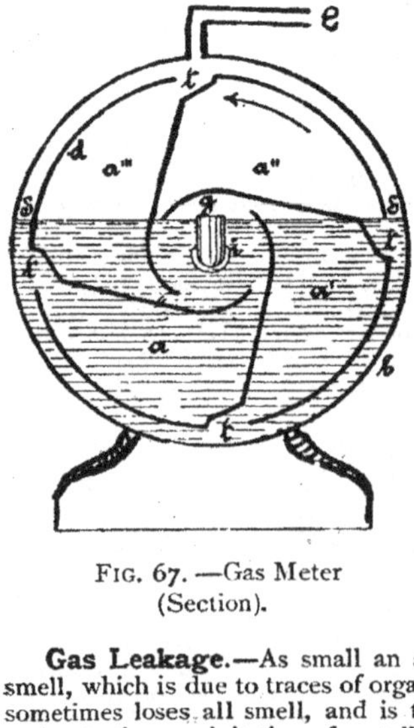

FIG. 67.—Gas Meter (Section).

The quantity of gas used is measured by means of a "gas meter," which usually consists of a vessel *b*, half filled with water (or better, glycerol water to prevent freezing in winter time), and in which a drum rotates, which is divided into four compartments, *a*, *a'*, *a''*, *a'''*, as shown in **Fig. 67.** The gas enters the drum at *g*, and escapes through *t* and *e*. As the gas gradually enters at *g*, it forces the drum to rotate in the direction of the arrow, and the motion being communicated by means of the axle on which the inner drum rotates to a clock-work arrangement, gives the amount of gas which has passed through. The level of the liquid in the gasometer must be kept the same (by the addition of water) if the readings are to remain the same. Dry gas meters are also in use, consisting of two leather bags, one of which fills as the other empties; also automatic meters, a penny slipped in causing the delivery of a certain fixed volume of gas.

The pressure of the gas in the gasometer usually amounts to 70-80 mm., and passes out at the burners at a pressure of 25-30 mm., although now higher pressures than this are being used.

Gas Leakage.—As small an amount as 0.02 per cent. volume of gas can be detected by its smell, which is due to traces of organic N and S compounds. When gas filters up through earth it sometimes loses all smell, and is then detected by palladium chloride paper, which turns black owing to the precipitating of metallic Pd by the CO. Long breathing of air containing only 2 per cent. of coal gas (=0.1-0.2 volume per cent. of CO), causes poisoning. 1 volume gas + 6 volumes air is a maximum explosive mixture. Mixtures of *less* than 5 per cent. gas or *more* than 30 per cent. are non-explosive.

CARBURETTED WATER GAS

Water gas is the general name for the mixture of gases obtained by the decomposition of steam by means of incandescent carbon. In practice either anthracite or coke is used, and the gases consist of a mixture of hydrogen, carbon monoxide, and carbon dioxide. The reaction is endothermic, so that heat must be applied or the incandescent carbon soon becomes cooled, when the reaction ceases.

The mixture of gas burns with an intensely hot flame which is non-luminous. When it is to be used for illuminating purposes it must, therefore, be enriched by adding volatile hydrocarbons obtained by "cracking" or heating to a high temperature crude petroleum products, which burn with a highly luminous flame. A second reason for adding these hydrocarbons is to render the gas odorous, for the mixture is very poisonous, and being without smell it is highly dangerous unless thus mixed with some gas whose powerful odour renders it easy of detection. This mixture is known as carburetted water gas.

It is usual to prepare the volatile hydrocarbons (oil gas) and the water gas simultaneously in the same plant. The gases are mixed as soon as formed, and are then generally submitted to a high temperature in order to prevent the subsequent deposition of the hydro-carbons, which before this final heating are not strictly permanent gases.

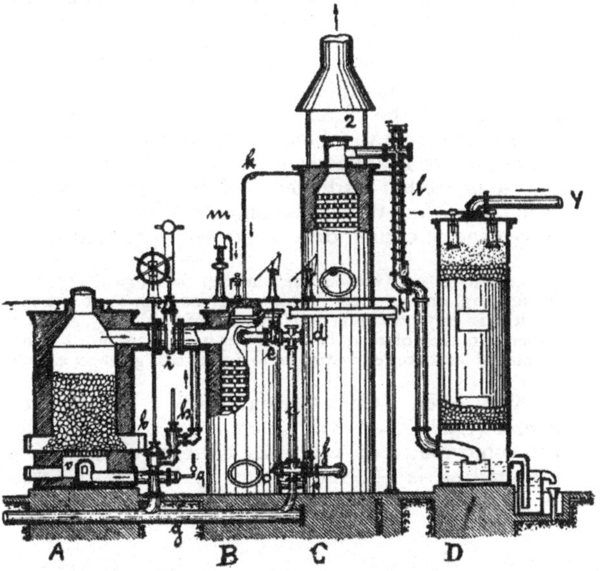

FIG. 68.—Humphrey & Glasgow's Carburetted Water Gas Plant.

The forms of plant for the manufacture of carburetted water gas are very numerous and variable. They may, however, be broadly classed under two heads: (1) The continuous system, in which steam is passed uninterruptedly into a chamber containing the anthracite or coke, which is maintained at the requisite temperature by external furnaces. This system secures a continuous supply of gas but is not economical, as a considerable portion of the heat is lost in passing through the walls of the chamber. It is very little used. (2) The intermittent system, in which steam and air are alternately blown into a mass of incandescent fuel, usually in opposite directions. The steam is decomposed by the hot carbon with consequent cooling of the mass, which is then raised to the required temperature by means of the air blast. All the common forms of plant at present in use belong to this system. They are all arranged so as to secure the production and maintenance of a sufficiently high temperature (about 1,000° C.) in the generator, the presence of a sufficient depth of fuel, and a means of regulating the supply of steam and air.

Fig. 68 shows Humphrey & Glasgow's apparatus. The generator A is a steel shell lined with firebricks. It is charged with anthracite or coke which is ignited and submitted to an air-blast, which is forced in through the tube *g* and enters at *v*. Part of the fuel burns to CO_2, which, passing up through the hot coke, burns to producer gas ($CO_2 + C = 2CO$). This passes away through *i* and enters the *carburettor* B, passes down this and then enters the *superheater* C at the bottom by means of the tube *p*. Both carburettor and superheater are steel shells lined with fire-brick and filled with brick checker-work; simultaneously with the entrance of the gas into them a stream of air from *g* is blown in through the side tubes *h*, *f*, *d*, *e*, causing the CO to burn and thus heat the brickwork in B and C to a red heat; the gas then passes out of the furnace through the opened stack valve Z, escaping into the air as CO_2. In the Lowe process, as practised in the U.S.A., the production of CO and the subsequent blowing in of air into B and C to burn it is avoided by blowing air very rapidly through the producer A, so that a sufficiency of oxygen is present all the time, and only very hot CO_2 passes away into B and C and heats them to the required temperature. When the proper temperature of the different parts of the apparatus is obtained the air blasts are shut off, beginning with that of the superheater C, and the stack valve Z is closed. Steam is then blown into the generator A through the tube *q*, and is decomposed by the hot carbon in A according to the equation $H_2O + C = CO + H_2$. It then passes into the carburettor as water gas. At the same moment the oil is introduced into the carburettor B, being pumped in through the tube *k*, and falling on the red-hot bricks in B is gasified, its vapours mingling with the water gas and passing on to the superheater C, where the oil vapours are permanently gasified, and thence through *l* (where the incoming oil is preheated by the hot gas) into the washer D (consisting of a tower filled with coke down which water trickles), where it is washed and cooled, finally passing away at Y to the purifiers and the gas-holder. When the temperature of the mass of coke becomes too low the steam is shut off, the stack valve Z opened, and the air-blast readmitted.

The gas thus prepared contains rather less than 3 per cent. of carbon dioxide, and about 0.2 per cent. of sulphuretted hydrogen. It is practically free from ammonia and the other impurities commonly found in coal gas. The purification of the gas is, therefore, a comparatively simple matter: it is simply washed and passed through two "lime" and two "ferric oxide" purifiers in series. Sometimes a fifth purifier is added as a check to ensure the complete removal of carbon dioxide and sulphuretted hydrogen. The purifiers are similar to those used in coal-gas purification, but those containing the lime must be large on account of the high percentage of carbon dioxide present in the gas.

Since water gas (without the admixture of oil gas) possesses a calorific value of 2,500 calories per cubic metre, and pure oil gas one of 8,600 calories, it is clear that by gasifying a suitable amount of oil it is easy to obtain a gas of calorific value = 5,000 K and of 16 candle-power illuminating power, and even of 20-25 candle-power (with more oil). Hence the addition of only 10-20 per cent. of this rich carburetted water gas to a poor coal gas very essentially improves it (see pp. 119, 126).

Very often "Cold Carburetting" is resorted to in Germany, the agent being benzol (C_6H_6, B.P. 81°, which at 10° C. can exist as vapour to the extent of 200 g. per cubic metre in water gas, corresponding to 6 per cent. by volume), and the carburetting is effected merely by leading the water gas through a tower down which a rain of warm benzol is falling.

OIL GAS

Just as coal gas is prepared by the destructive distillation of bituminous coal, so oil gas is prepared by the destructive distillation of mineral oil. Several processes are in use for the manufacture of this gas, by far the most important of which is the "**Pintsch**" **process.** The gas is formed by the vaporisation of mineral oil. The most important oils used for this purpose are "Mineral seal" oil, Scotch shale oil, and Russian "Solar" oil.

The manufacture of oil gas on the *Pintsch* system is carried out in gasworks somewhat similar to those used for the production of coal gas, but on a smaller scale. The main building is usually divided into three divisions, the *retort chamber*, the *purifying chamber*, and the *engine-room*. The gas-holder is placed at one end of the building, and the high-pressure storeholders at the other.

Fig. 69 shows a typical Pintsch oil gasworks. The oil is placed in a large tank sunk in the ground outside the building, from which it is pumped as required into a small cistern placed over the retort bench. A pipe leading from the cistern conveys the oil to the back ends of the upper retorts, the rate of flow being controlled by means of suitable valves. These retorts, which may be of cast iron or fireclay, and which are of ◘-shaped section, are set horizontally in pairs one above the other and connected by cast-iron mouthpieces at one end. They are heated by means of the firebrick furnaces A. The upper retorts contain sheet-iron trays, which are sometimes filled with coke, along which the oil flows to the front of the retorts, whence it descends through pipes to the lower retorts. The oil, which has now become changed into a thick yellow vapour, passes from the back of the lower retorts down the descension pipes into the *hydraulic main*, and thence through the *condensers* B, which are usually plain vertical iron cylinders, and the *washer* C, into the *purifiers* D. The latter are somewhat similar to those used in coal gasworks: the lower portion contains water, above which are wicker trays (usually three in number), covered by a layer of about 2 in. in depth of lime, or a mixture of lime and sawdust. The gas is caused to pass through the water and over the lime and thence through the meter E into the gas-holder F. It is finally pumped from this holder by means of the compressing engines G into the high-pressure storeholders H, where it is kept at a pressure of 10 atmospheres for a short time in order to enable some of the heavier hydrocarbons to settle out as a liquid at the bottom of the holders, after which it is ready for use. The tarry matter deposited in the hydraulic main, condensers, etc., is conveyed by means of pipes into the tar tank J, the quantity of tar thus obtained being usually about 5 to 6 gals. per 1,000 cub. ft. of gas.

Pope's system of making oil gas is very similar to the above, but the oil is passed in a fine spray into the lower retort whence the vapours pass into the upper one for fixing. There are no trays in either retort.

Keith's system is designed to give a large volume of gas, though at the expense of the quality. In order to effect this the retorts are flattened towards the middle of their length, thus promoting the contact of the oil vapours with the hot surfaces of the retort at this part.

In *Tatham's* process about 15 per cent. of oxygen is added to the oil gas in order to increase the candle-power.

Oil gas consists of a mixture of CH_4, C_2H_4, C_2H_2, C_6H_6, etc., possessing a very high illuminating and heating power; but it will only burn smokeless from burners with very narrow slit orifices. Since oil gas can be easily prepared in small plants and requires only a very simple purification, it was formerly much used for lighting factories or houses remote from gasworks, but has now been displaced by electric light. The principal users are the railway companies, who deliver it compressed to 10 atmospheres at the stations, and it is stored in the iron holders under the railway carriages under 6-8 atmospheres pressure. Until recently acetylene gas (25 per cent.) was mixed with it to improve its illuminating power, but now pure oil gas is again used, being burnt in special inverted incandescent mantle gas-burners, giving cheaper light than electric light.

The great advantage of oil gas over coal gas is its power of retaining a large portion of its luminosity when compressed. Both gases under these conditions deposit liquid hydrocarbons, but whilst coal gas loses nearly the whole of its illuminating power oil gas only loses a comparatively small portion. It is therefore particularly suitable for use wherever it is necessary or desirable to store the gas under pressure, as in railway carriages, steamboats, lighthouses, and buoys.

AIR GAS

Air gas is the most serious rival with which acetylene has so far had to compete. It is economical, safe, and clean, the light is excellent, at least when used with incandescent mantles, and there is no troublesome by-product to be got rid of. It consists of a mixture of air and the vapour of some volatile hydrocarbon, usually petrol or carburine, but sometimes benzol. The mixture may contain too little hydrocarbon to form an explosive mixture or too much, *i.e.*, it must either contain not more than 2 per cent. or not less than 7 per cent. of petrol vapour. In the first case the gas is always used with mantles as it burns with a non-luminous flame, in the second it may be so used but is usually burnt in self-luminous burners. The plant for the production of this gas is very simple. There are several types, differing in detail and more or less in efficiency and safety.

A good type is that supplied by the Non-Explosive Gas Company (see Fig. 70).

A small hot-air engine M heated by a Bunsen burner, which burns the air gas itself, drives a

Root's blower L and thus forces a current of air through the carburettor F, where it mixes with the hydrocarbon vapour. The hydrocarbon is placed in a tank Z beneath the carburettor. Inside this tank is a horizontal axle bearing six—or in the most modern forms twelve—arms, each of which carries a tipping bucket C which empties the hydrocarbon into a feed pipe E leading into the carburettor F. The latter is a large pipe about 3 in. in diameter, inside which are overlapping ledges. The petrol vapour flows over these, and, owing to the large surface exposed, quickly volatilises. The air gas passes from the carburettor F, up the tube P, through a meter K, into the holder, which is of iron and is always of small size—about 18 in. high and 16 in. in diameter for plants supplying less than a hundred lights. The axle of the drum of the meter is extended through the front and is coupled up with the axle carrying the tipping buckets in the petrol tank, so that the rate at which the latter revolves, and therefore the rate at which the hydrocarbon is supplied to the carburettor, is regulated by the rate at which the drum of the meter is turning, which in its turn depends upon the rate at which air passes through the carburettor, and the resulting gas passes through the meter. By this means the percentage of hydrocarbon vapour is kept constant.

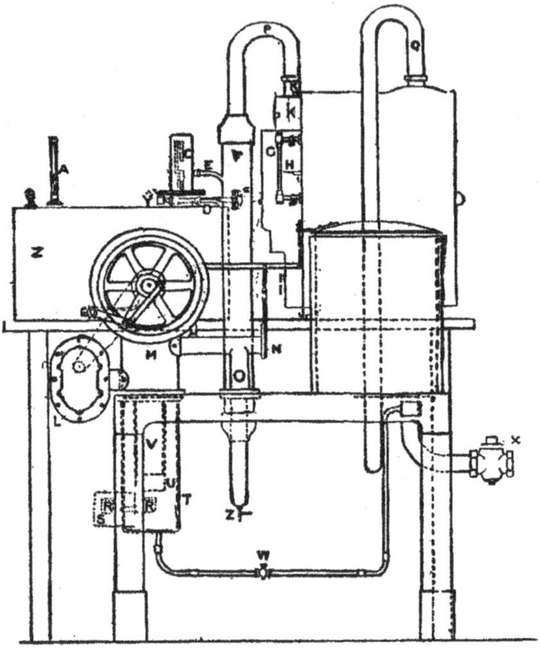

FIG. 70.—Plant for Air Gas.
(The Non-Explosive Air Gas Co. Ltd., London.)

The gas is consumed in burners with small inverted mantles. The oil usually employed is petroleum ether, B.P. 35°-60° C.

NATURAL GAS

In some parts of the world, especially in America, immense stores of natural gas under great pressure have been obtained, generally when boring for oil. This gas is conveyed in pipes to the various towns where it is required for use, chiefly for metallurgical purposes. The composition is variable.

CHAPTER XII

Ozone

CHAPTER XII

INDUSTRIAL OZONE

By E. Jobling, A.R.C.Sc., B.Sc., F.C.S.

The importance of ozone, or *triatomic* oxygen, depends upon the extreme readiness with which it parts with its third atom of oxygen, particularly when in contact with oxidisable substances, to give ordinary or *diatomic* oxygen. Ozone is therefore an energetic oxidising agent, and as such has met with extensive industrial application—a circumstance which is to be attributed as much to the non-formation of any troublesome after-product as to the nascency of the liberated oxygen.

Though always prepared and used in a very rarefied form, ozone may be condensed from the gas containing it as a deep blue liquid of B.P. −119° C. (the pure gas is itself blue), which is dangerous to handle on account of its tendency to explosive decomposition. It has a characteristic odour, perceptible even in extreme dilution with air (1 part O_3 in 1,000,000 air). Gaseous ozone in the rarefied condition spontaneously decomposes, the rate of decomposition depending both on the temperature and the concentration. At 270° C. ozone cannot exist; at ordinary temperatures, though, air charged with ozone is fairly stable. Like air and oxygen, ozone is almost insoluble in water.

1. **Thermal Method**.—Since heat is absorbed by the formation of ozone from oxygen—

$$3O_2 = 2O_3 - 59{,}000 \text{ cals.},$$

it follows that increase of temperature is favourable to ozone formation. Nernst utilises this as the basis of a process in which a current of carefully dried air is brought into contact with an electrically-heated non-metallic body, *e.g.*, a Nernst filament (D.R.P., 195,985 of 1906). To minimise the decomposing effect of temperature upon the product, the velocity of the air must be such that it is only momentarily exposed to intense heat and is then suddenly cooled. The method, however, is not a practicable one owing to the low ozone concentration developed. The most productive thermal method appears to be that of immersing a glowing Nernst filament into liquid oxygen (F. Fischer).

2. **Electrolytic Method.**—If, in the electrolysis of aqueous liquids, the conditions are arranged for a very high current density at the anode, the oxygen liberated there is found to be ozonised. Sulphuric acid has been shown to be the best adapted for the purpose (Kremann). To obtain a good yield, as well as to prevent destruction of the anode, the latter must be cooled, either by rotation or by means of internally flowing water or salt solution (F. Fischer). The yield is further increased by partly enclosing the anodes with non-conducting material so as to leave only long thin lines of the platinum or other metal exposed (F. Fischer).

For the above electrolysis, direct current of course is employed, but in a new method for large scale ozone production recently described by Archibald and Wurtemberg, both direct and alternating current are employed in the electrolysis of dilute sulphuric acid. The alternating current is found to act as a depolariser, leading to highly increased yields of ozone. In one instance, the production of ozone was 300 times greater than with direct current only. The higher the frequency of the alternating current, the greater the yield of ozone.

3. **Photo-chemical Method.**—Lenard discovered that air or oxygen is "activated" by ultra-violet light. For the production of this light of very small wave-length, mercury vapour or Uriol lamps are employed, the lamp itself being entirely or in part of quartz, since quartz, unlike glass, is transparent to ultra-violet rays. Potter (American Patent, 845,965) takes advantage of these facts for the construction of an ozoniser in which a quartz mercury-vapour lamp is surrounded by a sheathing of ordinary glass, between which and the quartz of the lamp air is spirally conducted.

4. **Electrical Method.**—For industrial purposes, the only practicable method of producing ozone is that of passing dry air through a chamber where it can be subjected to the influence of a silent high-tension electric discharge.

It is not certain whether the production of ozone by silent discharge is fundamentally an electrolytic phenomenon, dependent upon the electric stress developed, or whether it is to be attributed to the ultra-violet rays developed. The latter is the more probable (Harries), in which case the preceding photo-chemical method is only a modification of the present one and vice versa.

In its simplest form the apparatus employed is the well-known Siemens ozone tube, comprising two concentric cylinders of glass connected to the poles of an induction coil, and between which passes a current of air. Commercial ozonisers are constructed on the same principle, the different systems being distinguished chiefly by the arrangement and material of the electrodes, and the presence or absence of a solid dielectric.

The tension of the current employed depends upon the thickness of the dielectric (if any), the nature of the electrodes and their disposition, and varies between 10,000 and 90,000 volts. The higher the tension, the greater the output of ozone, the latter being theoretically proportional to the potential difference. A high frequency, at least 100 alternations per second, is found necessary.

A solid dielectric between the electrodes is now generally considered a more efficient arrangement than reliance only upon the air or oxygen passing between the plates, as it both increases the regularity of discharge, and diminishes the tendency to sparking.

The current has a deozonising effect which increases with the concentration of the ozone. The concentration must therefore be kept low. There is a limit to which it may be pushed, but this maximum is far above the workable concentration level, since the energy then demanded is out of all proportion to that required at lower concentration.

In most of the technical apparatus afterwards to be described, the concentration of ozone in grams per cubic metre of ozonised air amounts to an average of about two. At this concentration, 30-60 gr. of ozone per kilowatt-hour are produced at a cost of ½d. to 1d. For higher concentrations, implying a slower air current, the yield is less, and vice versa. This efficiency of production is still far distant from that theoretically obtainable.

Heat and humidity are unfavourable to the generation of ozone. The influence of the former in causing a reversion of the ozone to oxygen has already been remarked. To minimise it, the electrodes are cooled by any suitable means, though the ideal generator would naturally be that in which non-production of heat is attained rather than its elimination by cooling means. To prevent the very considerable loss which moisture introduces, the air delivered to an ozoniser is always first dried by contact with a hygroscopic solid or by refrigeration. Dust is undesirable, so that the incoming air is usually also filtered.

Sparking in the discharge must be rigorously prevented, for it leads to the formation of oxides of nitrogen which are harmful to the apparatus, and usually detrimental to the quality of the ozonised air produced.

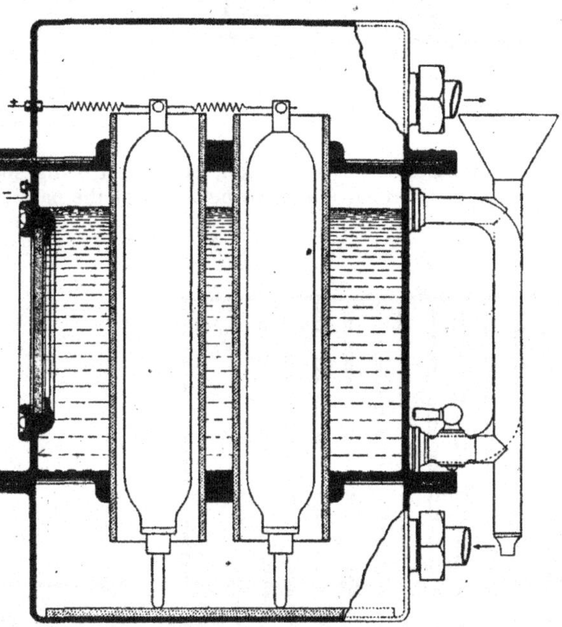

FIG. 71.—Siemens and Halske Process.

The following include the most important of the ozonising plants now in vogue:—

1. **Siemens and Halske Process.**—As shown in Fig. 71, each ozone tube comprises a cylinder of glass surrounding a cylinder of aluminium, and six or eight of these elements are usually mounted, vertically or horizontally, in a metal case through which water flows for cooling purposes. The cylinders are connected to the A.C. mains, the glass cylinders being put to earth through the surrounding water and case to reduce the risk of shock. Air enters the lower chamber, passes through the ozone tubes, and is drawn off as ozonised air from the upper chamber. As viewed through a glass window, the discharge is seen to be blue, and to enable the attendant, therefore, to see at a glance whether the ozonisers are working properly, they are kept in a dark room.

2. **Otto System.**—In the earlier form of this ozoniser (Fig. 72) one electrode consisted of a fixed iron cylinder, while the other comprised a large number of aluminium discs arranged perpendicularly to a central axle of the cylinder. The

rotating discs were provided with two opposite segmental openings, those of each disc being staggered with respect to adjacent discs, whereby any spark that was formed was drawn out and ruptured by the rotation of the discs. No dielectric was employed.

In the latest form, however (see Fig. 73), glass is employed as a dielectric, and the ozoniser now consists of an outer casing in which are arranged a number of sheets

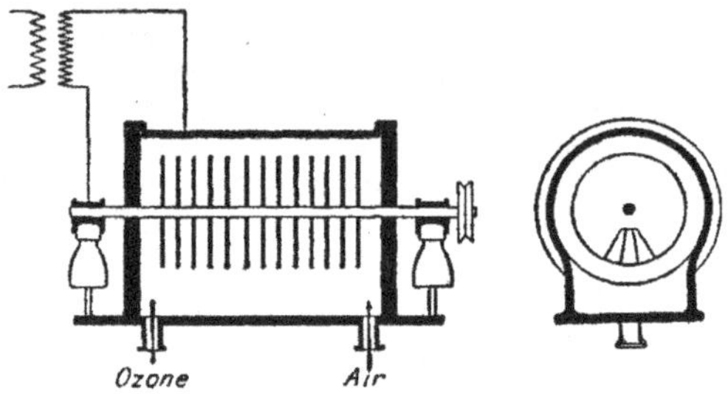

FIG. 72.—Otto System. (Old Form.)

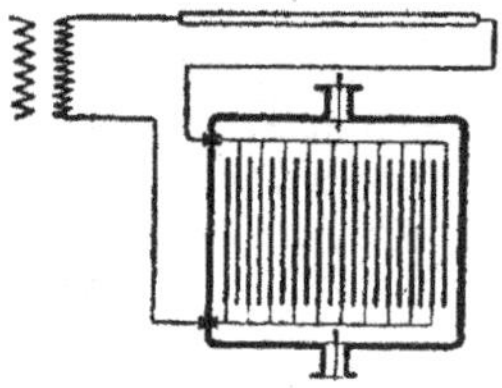

FIG. 73.—Otto System. (New Form.)

of glass coated on alternate sides with tin-foil, each sheet being separated by a strip of insulating material so as to leave a space between for the passage of the air.

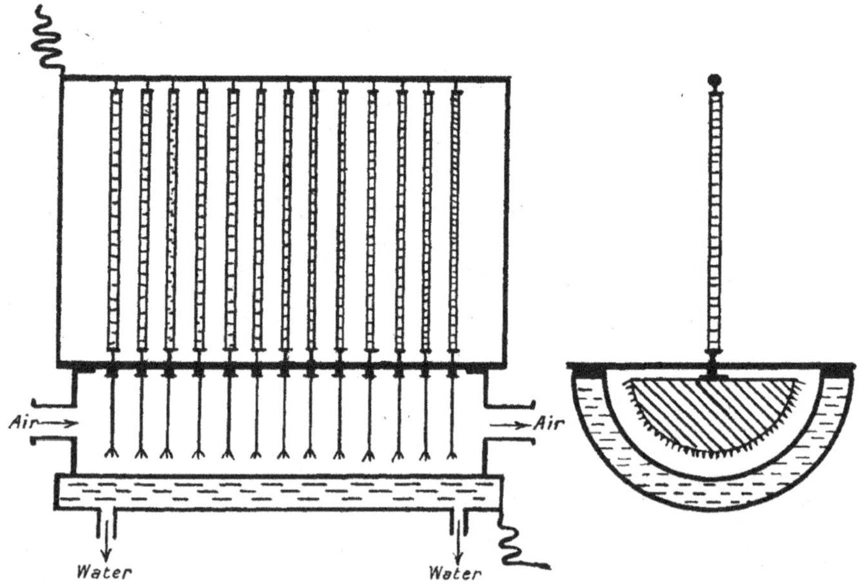

FIG. 74.—De Vrise System.

Longitudinal Section. Cross Section.

3. **Tindal-de Vrise System.**—No solid dielectric is here employed. Each ozoniser (Fig. 74) comprises a horizontal semicylindrical metallic trough, fitted hermetically with a glass cover and provided externally with a water jacket. Semi-discs of metal, having serrated edges, are suspended from the cover at short intervals and form one electrode. The metal trough is earthed and forms the other electrode. To prevent sparking, a series of high liquid resistances, consisting of tubes filled with glycerine and water, are arranged in the circuit. The silent discharge takes place between the semicircular high-tension poles and the water-cooled inner surface

of the trough, and between the poles the air to be ozonised is circulated. Five or six such cells are arranged in series.

A similar ozoniser, due to Tindal, is shown in Fig. 75. The arrangement is self-explanatory from the diagram given.

4. **Abraham-Marmier System.**—In this ozoniser the discharge surfaces consist of glass plates whose outer surfaces are cooled by water circulating in the

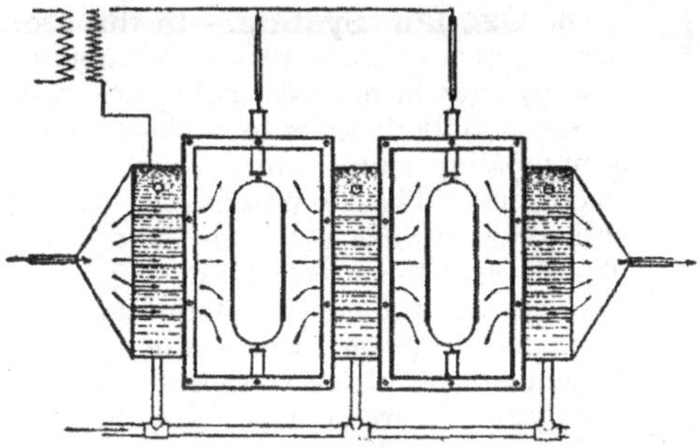

FIG. 75.—Tindal's System.

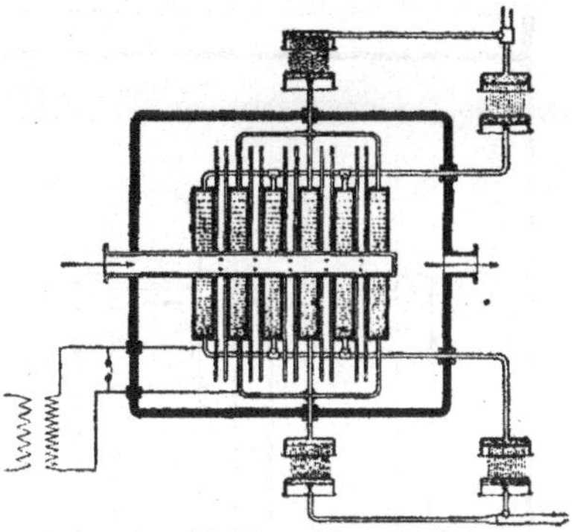

FIG. 76.—Abraham-Marmier System.

surrounding air-tight metal tank. Since the water serves as a conductor, being in contact with the high-tension poles, a high resistance, in the form of a number of water showers, is employed to prevent short-circuiting through the water (Fig. 76).

5. **Vosmaer System.**—Dielectrics are dispensed with in this system also. The ozoniser comprises a series of parallel tubes, each of which contains one flat

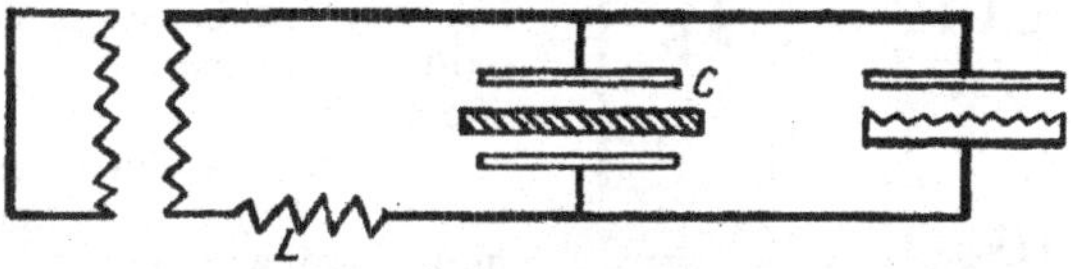

FIG. 77.—Vosmaer System.

and one saw-edged electrode held at a fixed distance apart by porcelain insulators. Sparking is suppressed by the insertion of a condenser C and induction coil L in the circuit (see Fig. 77).

6. **Howard-Bridge System.**—The discharge space in this case is bounded by two concentric cylinders, the outer of glass lined with aluminium and the inner

FIG. 78.—Howard-Bridge Ozoniser.

of metal pierced by a large number of holes for the passage of the air. Concentration of the discharge is effected by raising the rims of the holes above the level of the inner cylinders (see Fig. 78).

7. **Gérard System.**—This ozoniser consists essentially of two concentric glass cylinders of about a metre in length, which thus form a double dielectric, whilst the electrodes are formed of metallic sheathings affixed to the outer surface of the outer tube and to the inner surface of the inner tube. Air is passed down the

inner tube (see Fig. 79), rises between the dielectrics, and escapes charged with ozone. Several, usually ten, of these elements are arranged vertically in a refrigerator consisting of a tank containing oil, and each element is surrounded by a wide cylinder of zinc, considered by the inventor both to increase the circulation of the oil, and, by acting as a condenser, to promote the formation of ozone. The advantage of oil over water as a cooler lies in its insulating properties, for on the accidental breakage of a tube it prevents short-circuiting as it cuts the tube off from the remainder.

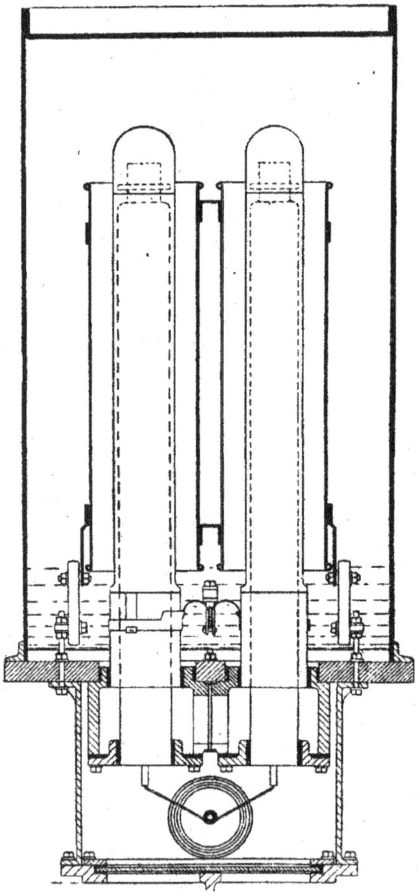

FIG. 79.—Gérard System.

8. **Ozonair System.**—In this ozoniser a series of mica plates are mounted side by side in a case, each plate being covered on both sides by a sheet of aluminium-alloy gauze which serves as the electrodes. The air passes into the box and between the plates in a direction parallel with them. The use of gauze is claimed to promote the formation of a sparkless discharge, whilst the open arrangement of plates is said to be sufficiently preventive of heating without the aid of water-cooling.

Detection and Determination of Ozone

Detection.—Ozone is most readily detected by the "blueing" of starch-potassium iodide paper. Unfortunately, however, other substances, *e.g.*, nitrous oxides, chlorine, etc., produce the same effect.

A more conclusive test is obtained by using a litmus or phenolphthalein paper moistened with neutral potassium iodide solution. Chlorine, nitrous acid, etc., if present, form the neutral salt, and therefore give no reaction. Ammonia is the only likely substance to disturb the reaction, and this can be detected by the use of a separate litmus paper. Ozone oxidises the potassium iodide to caustic potash and turns the litmus blue.

The most sensitive qualitative test is obtained by the use of the so-called "tetramethyl-base" paper, *i.e.*, paper soaked in an alcoholic solution of tetramethyl-di-*p*-aminophenylmethane. The moist paper is turned violet by ozone, straw yellow by nitrous oxide, deep blue by chlorine or bromine, and is not affected by hydrogen peroxide (Clement, *Ann.*, [4], **14**, 334, 1902).

Estimation.—The standard industrial method is that of absorption by neutral potassium iodide solution, and after acidifying, by titration in the usual way with sodium thiosulphate and starch.

Satisfactory results have also been obtained by Ladenburg (*B.*, **36**, 115, 1903) for a method in which the ozone was passed into sodium hydrogen sulphite and then titrated. In either of the above cases if the presence of other oxidising agents is suspected, a duplicate estimation should be made after the gas has been heated above 270°, when the difference in the results gives the amount of ozone.

For the separation from small quantities of the usual impurity, namely, nitrous oxides, the gaseous mixture should be passed into liquid air when the ozone dissolves, whilst the nitrous oxides separate as blue flocks. On filtration and evaporation both constituents are determinable (Fischer and Marx, *B.*, **39**, 2556, 1906).

Application of Ozone

Water Purification.—Ozone is known to exert a destructive action upon many of the micro-organisms found in water, particularly those which are deleterious to health, and in consequence of this, it has been utilised on a very large scale for purposes of water sterilisation. The method simply amounts to a cold combustion of the bacteria and organic matter present.

The present methods of water purification[1] comprise:—

1. The use of antiseptic chemicals, usually chlorine or bromine in some form.
2. Sand filtration.
3. The employment of ultra-violet radiation.
4. Ozonisation.

Of these, sand filtration has long been relied upon, and is still employed for all ordinary waters. Treatment by chemicals has not found much favour industrially; whilst purification by ultra-violet light, though making rapid strides, is still in its infancy. Further consideration of these methods cannot be undertaken here.

Ozonisation finds its greatest usefulness in the purification of those waters, usually of superficial origin, which are more or less discoloured or contaminated. Its advantage lies in the fact that it leaves no objectionable residue, is non-toxic, its comparative insolubility prevents "over-dosing," and it does not attack the usual dissolved salts which give taste to potable water.

The salts of the alkalies and alkaline earths remain unattacked, as also do ferric salts, but ferrous and manganous salts are oxidised and precipitated. Organic matter is completely destroyed. To minimise the consumption of ozone, however, all organic and inorganic impurities must first be removed by mechanical or chemical means.

A modern ozonising plant consists of (1) ozonisers, (2) sand filter, (3) ozonising towers, and (4) a reservoir for the purified water. The various ozonisers have already been described. It only remains to deal with the various types of ozonising towers employed.

Since ozone is only very slightly soluble in water, satisfactory sterilisation is only obtained by the intimate contact of the two. This is accomplished in the following ways:—

1. **Surface Contact.**—In the Siemens and Halske tower, which illustrates this type of apparatus and of which a diagram is shown in Fig. 80, the previously filtered water is allowed to trickle down one or more towers filled with stones or gravel, while at the same time a current of ozonised air is forced through the tower in the opposite direction. The ozone not absorbed is returned to the ozoniser.

The Abraham-Marmier tower is of the same type.

This system is being replaced by a system which employs:—

2. **Emulsification.**—Otto's apparatus is the most important in this connection and constitutes a kind of injector in which the filtered water is forced into a small chamber provided with holes through which ozonised air is sucked in and churned into the water (Fig. 81). To complete the mixing action the apparatus terminates in a long piece of piping, 15 ft. or so in length, whereby the ozonised air is more readily absorbed by the water by being subjected to the increased pressure of the lower part—a method known as "self-contact."

Another form of ozonising apparatus, in which the ozonised air is sucked in by the flow of the water, is that of the Howard-Bridge system, shown in Fig. 82. In travelling through the first U-tube, partially used ozonised air is drawn in from the pipe *a*, the ozone absorbed, leaving the air to escape through a hole at *b*.

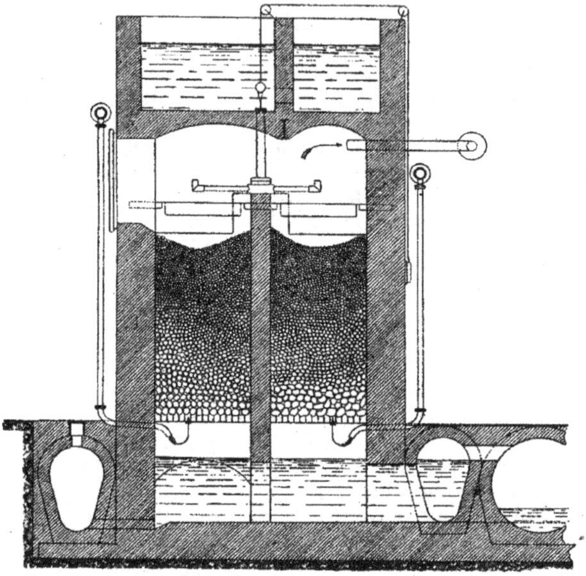

FIG. 80.—Siemen's and Halske Tower for Ozone Treatment of Water.

Fresh ozonised air is then sucked in from the ozoniser *c*, and accompanies the water along the second U-tube round the baffle plates and escapes into the reservoir. Any

FIG. 81.—Otto's Apparatus for Ozone Treatment of Water.

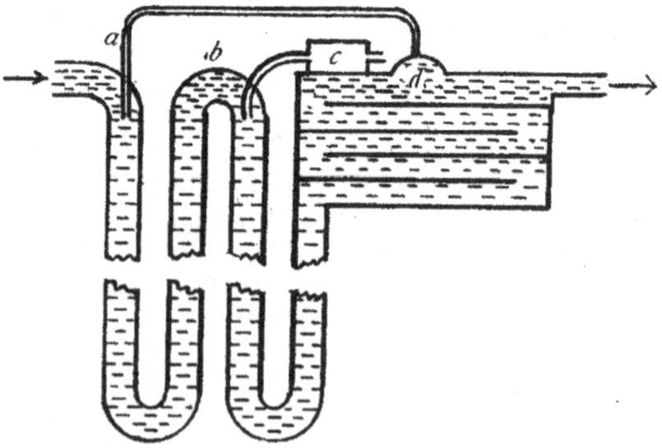

FIG. 82.—Howard-Bridge System for Treatment of Water.

unused ozonised air collects in the chamber *d* and is drawn back into the incoming water through the pipe *a*. This method effects considerable economy, for no power is required to compress the ozonised air into the steriliser. Incidentally it removes all difficulties connected with ozonised air pumps due to the lubricant being attacked by the ozone, for the pumps are dispensed with.

3. **Injection.**—Yet a third method is illustrated by the Siemens-de Vrise ozonising tower (Fig. 83). The towers are of iron, internally enamelled, and are divided into compartments by perforated lateral partitions of celluloid. The ozonised air is forced into the tower at the bottom, along with the water to be sterilised, and the ascending bubbles, by contact with the partitions, are suitably broken up for absorption of the ozone. By regulation of the speed of the water as it passes up the tower, the time of contact of the ozone with the water can be readily controlled.

The Vosmaer, Gérard, and Tindal towers are of the same type.

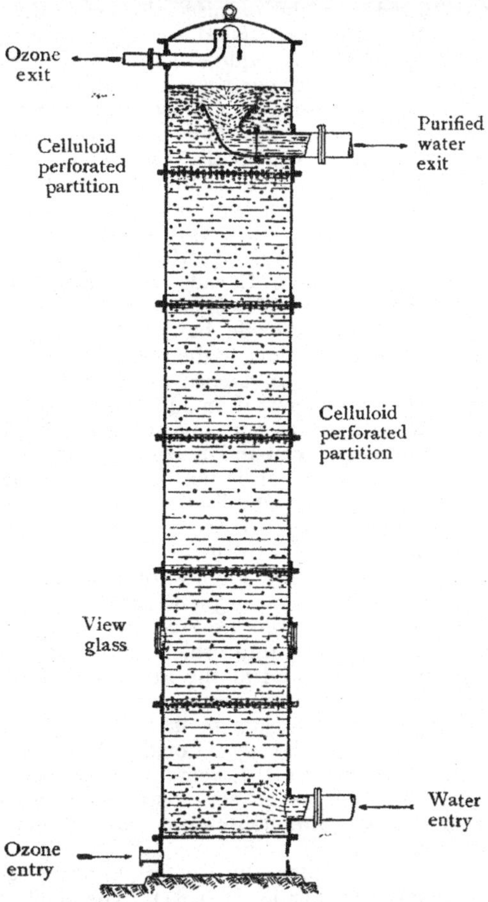

FIG. 83.—Siemens-de Vrise Water Ozonising Tower.

In the Ozonair type of apparatus the above three methods are combined in order to ensure satisfactory contact. The water is first atomised in the presence of ozonised air in the upper portion of the apparatus, and the particles of water are then allowed to fall upon a pile of glass spheres packed in a tower, where they meet an ascending stream of ozone. At the bottom of the tower the water falls into a tank, through which ozonised air is blown from nozzles arranged beneath the surface of the water, the last process being continued for any desired period.

In the case of all the ozonising towers above-mentioned, the ozonised water is discharged on to the top of a series of steps, down which it cascades into the storage tank and, by contact with the atmosphere, gives up its excess of ozone.

Recently, small ozonisers for domestic purposes have been placed upon the market. They are designed to fit on the ordinary water main, and are operated from the house supply, the ozonisers only functioning when the tap is turned on.

As to the amount of ozone required for complete sterilisation in water purification plant, it is not possible to say precisely. The amount varies with the water under consideration, and should be made the subject of preliminary research in each case. The average amount of ozone supplied in grams per cubic metre of water is about 2, though it may sink to ½ or rise to 8. The concentration of ozone in the ozonised air delivered from the ozoniser also lies in the neighbourhood of 2 g. per cubic metre, so that it may be stated broadly that a bulk of ozonised air equal to that of water to be purified is required. The cost of working, in the case of large installations, runs to about ½d. per 1,000 galls. of water treated.

On the Continent, water purification by ozonisation is largely practised, installations having been erected at Petrograd, Paris, Florence, Nice, and other large towns. The systems employed are usually the Siemens and Halske, de Vrise or Otto systems, or combinations of these. In England, the Ozonair apparatus has the widest application, but so far has not been adopted by any large public authority owing to the satisfactory purity of the water supply of most large English towns. The Vosmaer and Howard-Bridge systems are American in origin, and find their chief application in Canada and United States of America.

Air Purification.—The unpleasant effect of the air of crowded rooms is not due at all to an excess of carbon dioxide, but to the noxious organic exhalations which are given off from the skin and lungs of the people present. The introduction of ozone into such rooms is found to remove the source of the bad effects by oxidising these organic impurities, whilst if, in addition, a small excess of ozone be admitted, the air is rendered pleasant and invigorating. Bacteria, however, can hardly be removed, for this would demand a concentration of ozone which the human organism could not tolerate; but, if the incoming air be subjected to the influence of ozone, at least partial sterilisation can be effected. The costly complications introduced into heating arrangements by the necessity for hygienic ventilation are removed to a large extent by the use of an ozonising apparatus, which renders even the atmosphere of a closed room wholesome for a considerable time (see, however, *Journ. Ind. and Eng. Chem.*, **5**, 882 [1913]).

An air-purification plant usually comprises a filter screen through which the air is drawn and a mixing chamber into which ozone is introduced from the ozoniser to partially sterilise the air, and at the same time enrich it with a minute quantity of ozone (one part ozone to one million parts of air). From this chamber the air is distributed to the various parts of the building, etc., to be ventilated. Such installations have been erected by Ozonair, Ltd., in England and by Siemens & Halske in Germany, for the ventilation of large public buildings, factories, hospitals, etc. The method is also found to be particularly useful for sterilising the atmosphere of slaughter-houses and warehouses in which food of all descriptions is stored.

One of the tube railways in London, the Central London Railway, has recently been equipped with an ozonising installation (Fig. 84). The supply of fresh air is drawn from the outside through a filter F consisting of a gauze screen, where it is washed by atomised water, and its temperature somewhat reduced. A small proportion of the air is dried and passed into the ozoniser O, whence it is drawn into the mixing chamber M where it meets the remainder of the still moist air, and the whole is then passed on by ducts A to various parts of the tube. C shows the converter, T the transformer, and E the electric motor for driving the fan.

For household purposes, small portable ozonisers, comprising a transformer, motor, fan, and ozoniser, are upon the market.

Miscellaneous Applications.—In the brewing industry, the introduction of ozonised air into the fermenting chambers is found to have an invigorating effect upon the yeast, probably by reducing the number of bacterial and other organisms which prey upon it. Ozone is also employed with advantage in the sterilisation of casks, pipes, etc., and during bottling of the beer.

Ozone is employed, too, in the commercial manufacture of vanillin from iso-eugenol, and in the production of artificial camphor.

Its bleaching properties have found application in several directions, such as the decolorising of oils and fats, sugar, etc., and in the bleaching of delicate fabrics.

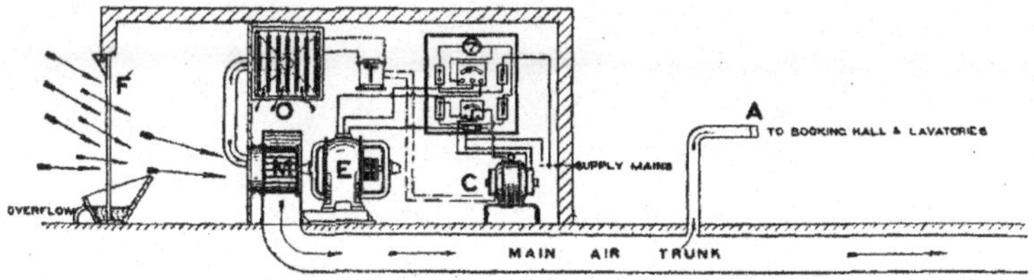

FIG. 84.—Ozonising Installation for Air Treatment of the Central London Railway.

It has been utilised, in addition, for the bleaching of flour, but the effect is now found to be produced by the oxides of nitrogen or hydrogen peroxide present and not by the ozone.

In the seasoning of timber, leather, tobacco, tea, etc., as also in the maturing of spirits, ozone finds useful application.

Acknowledgment.—The author desires to express thanks to Messrs Ozonair, Ltd., for permission to reproduce Fig. 84 from one of the several booklets on the subject which they are willing to supply.

SUBJECT INDEX

SAVE HARMLESS AGREEMENT

Because use of the information, instructions and materials discussed and shown in this book, document, electronic publication or other form of media is beyond our control, the purchaser or user agrees, without reservation to save Knowledge Publications Corporation, its agents, distributors, resellers, consultants, promoters, advisers or employees harmless from any and all claims, demands, actions, debts, liabilities, judgments, costs and attorney's fees arising out of, claimed on account of, or in any manner predicated upon loss of or damage to property, and injuries to or the death of any and all persons whatsoever, occurring in connection with or in any way incidental to or arising out of the purchase, sale, viewing, transfer of information and/or use of any and all property or information or in any manner caused or contributed to by the purchaser or the user or the viewer, their agents, servants, pets or employees, while in, upon, or about the sale or viewing or transfer of knowledge or use site on which the property is sold, offered for sale, or where the property or materials are used or viewed, discussed, communicated, or while going to or departing from such areas.

Laboratory work, scientific experiment, working with hydrogen, high temperatures, combustion gases as well as general chemistry with acids, bases and reactions and/or pressure vessels can be EXTREMELY DANGEROUS to use and possess or to be in the general vicinity of. To experiment with such methods and materials should be done ONLY by qualified and knowledgeable persons well versed and equipped to fabricate, handle, use and or store such materials. Inexperienced persons should first enlist the help of an experienced chemist, scientist or engineer before any activity thereof with such chemicals, methods and knowledge discussed in this media and other material distributed by KnowledgePublications Corporation or its agents. Be sure you know the laws, regulations and codes, local, county, state and federal, regarding the experimentation, construction and or use and storage of any equipment and or chemicals BEFORE you start. Safety must be practiced at all times. Users accept full responsibility and any and all liabilities associated in any way with the purchase and or use and viewing and communications of knowledge, information, methods and materials in this media.

Made in United States
North Haven, CT
11 November 2026

22197182R00088